T0290590

Taxing *Energy*

INDEPENDENT STUDIES IN POLITICAL ECONOMY

For further information on the Independent Institute's program and a catalog of publications, please contact:

The **INDEPENDENT INSTITUTE**

100 Swan Way, Oakland, CA 94621-1428

Taxing *Energy*

*Oil Severance Taxation
and the Economy*

Robert Deacon

Stephen DeCanio

H. E. Frech, III

M. Bruce Johnson

Foreword by Joseph P. Kalt

INDEPENDENT STUDIES IN
POLITICAL ECONOMY

THE INDEPENDENT INSTITUTE
100 Swan Way, Oakland, CA 94621-1428
510-632-1366 • Fax: 510-568-6040
ISBN: 0-945999-69-0   www.independent.org

Published in the United States of America 1990 by
Holmes & Meier Publishers, Inc.
30 Irving Place
New York, NY 10003

BOOK DESIGN BY DALE COTTON

The paper used in this publication meets the requirements
of the American National Standard for Permanence of Paper
for printed Library Materials, Z39.48-1984.

Library of Congress Cataloging-in-Publication Data

Taxing energy ; oil severance taxation and the economy / Robert Deacon
   . . . [et al.].
      p.   cm. — (Independent studies in political economy)
   Bibliography: p.
   Includes index.
   ISBN 0-8419-1179-7 (alk. paper)
   1. Petroleum—Taxation—California.   2. Gas, Natural—Taxation—
-California.   3. Severance tax—California.   4. Petroleum industry
and trade—California.   I. Deacon, Robert T.   II. Series.
HD9560.8.U53C28   1990
336.2'716—dc20   89-15390
                                                                      CIP

Manufactured in the United States of America

THE INDEPENDENT INSTITUTE is a tax-exempt, scholarly research and educational organization which sponsors comprehensive studies on the political economy of critical social and economic problems.

Contents

Taxing Energy

Acknowledgments

The initial research for this study was funded by the following firms: Atlantic Richfield Company, Chevron U.S.A., Inc., Exxon Company U.S.A., Getty Oil Company, Gulf Oil Corporation, Mobil Oil Corporation, Phillips Petroleum Company, Shell Oil Company, Sun Company, Inc., Texaco, Inc., and Union Oil Company of California. The conclusions presented here are, however, those of the authors alone and do not necessarily represent the views of the firms listed above.

Foreword

Mineral and energy deposits are often appealing tax targets for governments. They seem to represent perfect opportunities for collecting tax revenues without inducing the affected production facilities to move to another jurisdiction. "After all, the mine or the oil well can hardly migrate across state or local boundaries." But as this valuable book goes on to demonstrate, economic reality is not so simple.

Governments utilize taxes to pay for the services they provide and to redistribute income among constituencies. In pursuing these ends, governments have the capacity to significantly affect the productivity of the economy. Because the nets they throw over the economy are so wide, and because their taxes are felt as costs by those that must pay their levies, governments hardly need be neutral or benign when collecting their revenues. Poorly designed tax policy holds the potential for distorting the mix of outputs produced, relative to the mix that the consuming public desires; and when productive assets have the ability to flee to jurisdictions with lighter tax burdens, tax policy can constrain the absolute size of an economy. To the extent that these effects arise, the public's standard of living is held down and/or an excessively high price is paid for the attainment of redistributive goals.

What are good and what are poor tax policies? Economic principles generally prescribe the use of broad-based levies, such as consumption taxes, that leave an otherwise "level playing field" free of disproportionately taxed sinkholes. Taxing any one sector more heavily (on a percentage basis) than others can place that sector at a disadvantage when trying to attract customers that object to price increases or investors looking for profit opportunities.

The opportunity to tax immobile production facilities, such as oil wells, might seem to provide an exception to the prescription for broad-based taxes. Relatively high taxes on such facilities may have but minor depressing, distorting effects on supply, since "immobility" implies that the facilities have little choice except to stay put and produce. But this overlooks the flexibility of other variable factors, such as labor and capital, that are utilized in exploiting a fixed resource stock. Their use can be discouraged by excessive taxation, particularly when the industry can turn to supply sources in other jurisdictions—in the way that oil production can shift to Texas or the foreign sector if a producing state such as California, say, taxes too aggressively.

Notwithstanding the fact that taxes on mineral and energy resources may have negative effects on production, state governments in a federal system are particularly attracted to them. The *relative* immobility of a natural resource stock such as oil means that a relatively high percentage of any tax burden may be "exported" to other states. That is, the resource owners cannot flee with their oil to avoid paying taxes, and at least the corporate owners tend to be dispersed far beyond the borders of the taxing state. Consequently, the burden of state taxes on mineral and energy resources tends to be substantially external to the taxing state; and the externality can encourage states to overtax the resource sector. Economically, this incentive is limited only by the losses a state experiences as its resource taxes force the flight of the mobile capital and materials used in resource extraction and depress the demand for land and labor in the state.

At present, all major oil-producing states collect significant revenues on oil production through income, property, and severance taxes. California stands out for its failure to rely on any significant severance tax collections, depending instead on other taxes and royalties collected from state leases. This anomaly has prompted interest in some quarters for increasing levies on California oil producers, who appear to be more favorably treated than their counterparts in other states. A proposal to impose a severance tax of 6 percent on the gross revenues of California oil producers provides the immediate impetus for this study.

The authors, Professors Deacon, DeCanio, Frech, and Johnson, are noted economists and energy policy experts in their own right. Together, in *Taxing Energy*, they have produced a careful and accessible analysis of state energy taxation. Their factual findings are of particular relevance to California and other states in their consideration of severance taxes on oil production. It turns out, for example, that while California's tax burden on oil producers is slightly below average among the states, the combined revenues from taxes and royalties (expressed as a percent of the value of production) indicate that California is not "easy" on oil producers. In fact, California's oil tax system appears to be particularly well suited to its oil industry. Much of the production in the state is relatively high-cost and economically marginal. The state must tread carefully in taxing this production, lest it force it to be curtailed.

California already has one of the most efficient, least distorting energy tax systems in the nation (chapter 2). This is owing to the fact that California, unlike most other states that tax energy resource extraction, levies taxes primarily on net income, rather than gross revenues. This book reports that fully 90 percent of California's oil tax revenues comes from levies against net income, compared to 38 percent in Texas, 14 percent in Louisiana, and 13 percent in Oklahoma.

A tax on gross revenues forces producers to treat the tax as just another cost of production—a force that discourages incremental supply. A tax on net revenues, however, dampens this effect so long as producers are left with net, after-tax returns sufficient to leave them profitable. The severance tax being considered for California would be levied against gross revenue from oil production and thereby discourage development of California's resources. The authors here estimate that a 6 percent state severance tax ultimately would cause California to recover about 10 percent less oil than it would otherwise retrieve from its resource base. Along with this reduction would go a cut of 3–8 percent in the size of the workforce in California's resource extraction sector.

Perhaps even more important than its specific findings regarding the effects of an oil severance tax on the economy are this book's more general teachings on how to think about the systemic effects of tax policy. The lessons are important and carry across jurisdictions and economic sectors: while a particular tax target may appear to be captive and immobile, many of the complementary inputs to production, such as labor and materials, are variable in their use. When a tax policy aimed at garnering the rents associated with an immobile resource such as oil in the ground is implemented, labor and materials in the affected sector can be expected to be pulled out of, or never put into, production.

The result is an economic chain reaction: fewer workers carry home paychecks and fewer suppliers of materials earn their incomes in serving the taxed industry. The places where these "factor incomes" would otherwise be spent then experience some of the impact of the tax. Indeed, even the taxing government's attempts to raise overall revenue collections can be substantially thwarted as declines in factor incomes and the overall output of the economy are reduced by distortive taxation. This book carefully uses information on the input and output exchanges between economic sectors to illustrate the linkages in the economic chain that is a state's economy. The book reports, for example, that the induced decline in statewide labor demand upon imposition of an oil severance tax would be four to eight times larger than the workforce reduction predicted for the extraction industry. This suggests downward pressure on statewide wages and attendant reductions in the inducements to live in the state.

Finally, *Taxing Energy* is important for its illustrations of how to get from the blackboard of economic theory to the hard quantitative information that sound policy design requires. After first figuring out how a severance tax would affect the profit-seeking strategies of oil producers, the authors then combine detailed economic and geologic information on the relevant production processes to isolate the direct effects on California's oil industry. These direct effects then touch off secondary effects as they ripple through the state's economy, as captured in the book's input-output data. Following these steps is a most educational effort for the real-world tax analyst.

By pointing out the costs of taxing the oil industry, this book runs the risk of being interpreted as a defense of the status quo. But this book is not anti-taxes or pro–oil industry; it is only pro-analysis. Its policy message is clear: the public and its policy makers need to understand the overall consequences of attempts to find sources of government funding. Going after the apparently easy targets—the immobile assets of politically disfavored constituencies—is not free and may be downright counterproductive.

Joseph P. Kalt
John F. Kennedy School of Government
Harvard University

Chapter 1

Introduction

During the last five years the state of California has considered several proposals to raise the severance tax on crude oil from its current insignificant level to a rate of 6 percent. These proposals were seen as a partial remedy for the financial difficulties faced by the state during the early 1980s. This study addresses two economic issues relevant to the severance tax question: (1) Does California place an abnormally light tax burden on crude oil producers in the state? And (2) what effects would increasing the crude oil severance tax to 6 percent have on California's output, employment, and tax receipts?

Much of the sentiment to turn to the oil industry for additional tax revenue arose from a belief that state and local governments in California place an abnormally light financial burden on California oil producers. To clarify this question we obtained comparison data on actual government receipts in major oil producing states. During the 1979–80 and 1980–81 fiscal years, oil industry taxes in California (expressed as a fraction of gross sales) were similar in magnitude to those in Texas and Oklahoma but fell short of receipts in Alaska, Louisiana, and Wyoming. The average tax rate computed for California (7.3 percent of gross sales) was about one percentage point below the average for all states in the "lower 48." But the state of California also obtains income from crude oil royalties from production on state land. Considering both tax and royalty income, government revenues in California amounted to 13.4 percent of the value of nonfederal production in the state, which is below Alaska and Louisiana but well above Oklahoma, Texas, and Wyoming.

All taxes affect resource allocation; yet some cause fewer disruptions and inefficiencies in the functioning of the economy. California's current revenue system is unique among oil-producing states because it is primarily applied to net rather than gross incomes. By recognizing and

1

incorporating cost differences in the assessment of taxes and the collection of state royalties, this revenue system reduces disincentives that plague severance taxes and other levies on gross receipts. In this respect the revenue system now in place in California is well suited to the economically marginal nature of the state's crude oil resources.

In answer to the second question, our analysis of California crude oil resources indicates that a 6-percent severance tax would reduce the average remaining economic life of oil wells now producing in the state by about three years. Over the full lifetime of existing wells, the tax would reduce production by about 398 million barrels, or about 10.5 percent of the remaining production from those wells in the absence of the tax. Our results also indicate that a 6-percent severance tax would reduce the number of new wells drilled in the state by 230 to 280 per year (about 6.5 to 7.0 percent).

We estimate direct employment demand in fossil fuel extraction sectors would fall by an average of 1,700 to 2,600 positions over the 30-year period examined. When the indirect employment effects in other sectors are added, the total reduction in California employment demand would amount to 9,000 to 16,000 full-time equivalent positions.

The direct and indirect effects of the severance tax would also be reflected in lower levels of gross sales in the state. Estimated total reductions amount to $1.2 to 2.0 billion annually.

In addition to specific conclusions regarding the effects of a severance tax increase in California, this study develops methods and presents findings of more general interest. To our knowledge the estimated state-local tax burdens presented in our interstate tax comparisons cannot be found elsewhere in the literature. The average effective tax rates presented in these comparisons were computed from data on actual state and local tax receipts and should therefore portray accurately the pattern of tax instruments imposed in major oil-producing states.

Implicit in our analysis of a severance tax on production is a model of crude oil supply. When compared to other energy supply studies in the literature, the model we specify for California is notable in several respects. Most importantly, it uses a simple representation of the producers' profit-maximizing decision calculus to derive an explicit time path for the effects of net price changes on production from existing wells. This represents an advance over the ad hoc or purely empirical production time paths found in other supply models in the literature. Our analysis also makes use of highly disaggregated geologic data on production and reserves for 245 individual oil fields in the state.

Finally, our incorporation of input-output analysis with the statewide crude oil supply model appears novel and provides a vehicle for analyzing a variety of policy questions. By expressing the time pattern of production and drilling effects from the supply model as final demand changes for the input-output model, this approach enables the estimation of time paths of economic impacts. Our analysis focuses on the effects of a crude oil severance tax on employment demand and receipts from other taxes. The same approach could, however, be used to analyze other impacts of crude oil supply changes.

The remainder of this book is organized as follows. Chapter 2 presents our interstate comparisons of tax and government royalty burdens on crude oil production. The issue of tax exporting is addressed in chapter 3, and academic literature on this subject is reviewed to shed light on the likely degree to which a California crude oil severance tax would be exported. Chapter 4 provides a survey of the literature on oil and gas supply models and presents a brief analysis of relationships betwen the effects of tax changes and price changes on crude oil supply. Our crude oil supply model is presented in chapter 5 together with the simulated effects of a new 6-percent severance tax, and estimation of the employment and government revenue effects of the tax appear in chapter 6.

To make the analysis in this book accessible to the lay public, the concepts and methods found in the text are presented in an intuitive, nontechnical fasion. All mathematical and analytical details have therefore been reserved for appendices. Appendices *A*, *B*, *C*, and *D* provide methodological detail for chapters 2, 4, 5, and 6, respectively.

Chapter 2

State and Local Government Taxation of Crude Oil

2.1 Introduction

State and local governments gain a significant fraction of the value of the petroleum within their borders by taxation. The most important tax instruments are taxes imposed on the owners and operators of privately owned oil-producing properties in the form of severance or production taxes, property taxes, and corporate income taxes. The relative importance of these tax instruments varies widely among oil-producing states. In addition all state governments and several local governments in the largest oil-producing states own significant petroleum reserves. For such properties the governments receive bonus payments and royalties.

In the remainder of this chapter we provide estimates of the direct public revenue provided by crude oil resources, expressed as a portion of the gross value of total production, for all major oil-producing states in the United States. Our primary purpose is to provide comparisons of the ability of governments in various producing states to obtain, as revenue available for public purposes, shares of the value of oil resources found within their jurisdictions. In particular we seek to examine the nature of public revenue collection in California, and to compare and contrast it to systems found in other states.

It will become apparent that public sector shares of petroleum wealth, and the kinds of revenue instruments used to extract them, vary widely among major producing states. Although our aim in this chapter is primarily descriptive, the information furnished could provide a starting point for analysis of a variety of questions concerning the political economy of petroleum. For example, if effective tax rates and tax instruments vary widely across regions, then the effects of such taxes on resource allocation presumably will vary as well. In this regard it would be of interest to know if the tax systems in some states tend to be systematically more efficient than those found in other states. In a subsequent section of this chapter, we offer speculations on this question. A related question concerns the efficiency of tax systems found in various regions in relation to characteristics of the resources taxed. Some petroleum reserves are characterized by relatively high exploration and investment costs but moderate operating costs, while resources in other regions exhibit just the opposite characteristics. Since the structure of "optimal tax" systems for petroleum presumably varies with the nature of the resource, it would be interesting to know whether the observed variation in tax systems can be rationalized by regional differences in reservoir attributes. Finally it will be shown that overall tax rates are significantly higher in some states than in others. That is, governments in some oil-producing states apparently have been far more successful than those in other states in securing for public purposes a portion of the petroleum wealth in their jurisdictions. An attempt to explain why these differences exist might provide important insights into the political economy of petroleum.

2.2 Comparison States and Revenue Sources

The states included in our analysis, ranked according to the value of total oil production during 1981, are Texas, Louisiana, Alaska, California, Oklahoma, and Wyoming. Collectively these six states accounted for over 83 percent of all oil produced in the United States during 1981 (Independent Petroleum Association of America 1982). The major tax instruments incorporated in these comparisons are severance or production taxes, property taxes, corporate income taxes, and individual income taxes levied on royalty income. We also provide estimates of state government bonus and royalty payments from production on state-owned properties.

It is clear that given the tax instruments used by a particular government, the pattern of resource ownership will have a significant effect on state and local tax receipts. For example, the state of Alaska levies a substantial corporate income tax but no individual income tax. If significant amounts of oil were produced from properties owned by individuals and leased to producers under royalty arrangements, then such royalty oil would escape taxation. In Oklahoma, on the other hand, where individual tax rates are high relative to those levied on oil corporations, royalty oil is effectively taxed at a higher rate than oil for the operator's account. Federal ownership also has obvious effects on state revenues.

Constitutionally, federal properties are exempt from severance and property taxes imposed by lower levels of government. However, a state that levies a corporate income tax using the unitary approach can effectively tax part of the value of federal production if all or a portion of the operator's employment, assets, or sales falls within the state's jurisdiction. Indeed corporate income taxation is virtually the only way that state governments can appropriate a share of the value of resources found on federal properties. For these reasons it is important to keep in mind the distribution of petroleum resource ownership when estimating overall tax burdens.

2.3 The Representative Firm Approach to Comparing Tax Burdens

Comparisons of state and local tax burdens on petroleum might be approached in several ways. One approach that has been adopted in at least two recent studies relies primarily on the tax codes in various states and might be termed a *representative firm approach*. To implement this method one begins by characterizing a hypothetical oil-producing firm or oil property, an entity specified to be representative in terms of revenues, costs, reserves, capital equipment, and so forth, of operations found in the state in question. Statutory tax rates are then determined for taxes on corporate income, property, and production and used to estimate tax payments for the hypothetical operation. By performing such calculations for each comparison state, one obtains estimates of relative tax burdens for purposes of comparison.

Applications of the representative firm approach are California Assembly Office of Research (1981) and California Legislative Analyst (1982b), and scrutiny of these studies reveals some of the pitfalls inherent in this approach. The difficulty of applying this method stems in part from the diversity of tax practices in various states. Even if statutory tax rates or tax rate schedules were identical in a group of states, differences in the administrative details of how taxes are imposed make comparisons exceedingly difficult. For example, differences in property valuation methods, definitions of taxable property, assessment ratios, state depletion allowances, small-producer exemptions, and variances in application of unitary tax formulae are common, but seldom described in a way that permits one to make accurate allowances for them. Consider the case of property taxation. Two states, California and Texas, base property tax levies on the assessed value of production equipment and proved reserves. Wyoming, on the other hand, levies its property tax rate on the wellhead value of production in the preceding year. Louisiana taxes only production equipment but not reserves, while Oklahoma exempts both

Taxation of Crude Oil

reserves and equipment from property taxation. To complicate matters further, property tax rates are set at the local government level in several states, and rates vary from jurisdiction to jurisdiction.

An additional source of inaccuracy with the representative firm method arises from the diversity of production operations found in various regions. Crude oil resources in different states differ in such respects as gravity, location (for example, onshore versus offshore), climatic and environmental conditions (for example, Alaska versus the lower 48), depth of reservoirs, and the presence of associated gas. These differences are naturally reflected in the operating characteristics of firms producing in each state. In general no single hypothetical entity can represent accurately the nature of operations in all states. One might try to correct this problem by specifying different representative firms for each state, but to do so would strain comparability by introducing elements of judgment and uncertainty into any estimates eventually obtained.

The estimated tax burdens presented in table 2.1 exemplify the ambiguities that can result from this method of comparison. In the study cited two representative firms, corporations A and B, were characterized. Tax payments, expressed as fractions of each firm's gross sales, were estimated for each of six major producing states. The relatively wide range of average tax rates estimated for the two firms, particularly in Alaska, California, and Oklahoma, indicate the imprecision of this methodology. Moreover, as noted in section 2.4, estimated average tax rates reported in this table imply levels of state and local government receipts that differ markedly from actual tax revenues.

TABLE 2.1
Comparative Tax Burdens As Estimated from the Representative Firm Approach

State	Effective Tax Rate[a]	
	Corporation A	Corporation B
Alaska	17.5	14.4
California	6.1	3.2
Louisiana	14.7	12.7
Oklahoma	12.1	8.7
Texas	8.1	7.3
Wyoming	13.0	10.9

SOURCE: California Legislative Analyst (1982b).

[a] Expressed as percent of gross revenue.

2.4 Comparing Average Effective Tax Rates from Tax Revenue Data

A far simpler approach to the interstate comparison of tax burdens is taken in the remainder of this chapter. Rather than rely on tax codes and hypothetical firms, we simply compare actual petroleum-related tax collections in each of six major producing states, expressed as a fraction of the value of taxable production in each state. The resulting estimates represent average effective tax rates for each tax and each state. Most state and local governments regularly publish information on sources and magnitudes of public revenues. Since data from different states are not always reported in the same format, it is often necessary to adjust figures to permit comparability. The final results, however, reflect actual payments by petroleum producers. For the purpose of comparisons to be used to make policy regarding actual taxes, this is clearly preferable to judgments drawn from hypothetical examples.

The following analysis reports tax revenue comparisons for six states: Alaska, California, Louisiana, Oklahoma, Texas, and Wyoming. These are the six largest oil-producing states in the United States. Data were collected for five general revenue sources: severance taxes, state and local property taxes, corporate income and franchise taxes, individual income taxes on royalty income, and state lease royalties. The sources of these data are presented and discussed in appendix A. State tax revenue reports do not regularly publish information on individual income tax collections for royalty oil income. It was therefore necessary to estimate revenues from this source from information on state income tax codes and typical royalty rates.

When compiling data on state and local revenues, it was often necessary to make various adjustments to render them comparable. In some instances receipts for oil and gas were reported as a single sum. In such cases revenues were apportioned between oil and gas on the basis of the wellhead value of production of oil versus gas in the state. Generally, financial data were reported on a fiscal year basis. In cases where reporting was by calendar year instead, simple averages of observations for successive calendar years were used to approximate fiscal year receipts.

Four of the states examined levy corporate income taxes. Rates and allowances for percentage depletion are shown in table 2.2. All six of the states examined levy corporate franchise taxes, and in the following analysis these franchise taxes are lumped together with corporate income taxes and considered as charges against corporate income.

Of the states examined, Alaska and California report corporate income taxes by industry, and Texas reports franchise tax payments by industry.

TABLE 2.2
Corporate Income Tax Rates

State	Rate	Adjustments[a]
Alaska	1%–9.4%	Cost depletion only
California	9.6%	22% depletion allowed for small producers
Louisiana	4%–8%	38% depletion allowed for all producers
Oklahoma	4%	22% depletion allowed for all producers
Texas	0	
Wyoming	0	

SOURCE: See appendix A.

[a] Percentage depletion limited to 50 percent of net income in all states.

In all other cases (Louisiana, Oklahoma, and Wyoming), it was necessary to estimate the industrial breakdown of corporate income and/or franchise tax payments. The details of the method used are presented in appendix A. The estimation method takes into account allowances for percentage depletion and the consideration that, with unitary taxation, a fraction of the income earned from production on federal outer continental shelf (O.C.S.) leases is taxable in adjacent states if they represent the point of sale.

Sources and amounts of revenues collected in each state in 1979–80 and 1980–81 are presented in tables 2.3 and 2.4, respectively. With the exceptions of California and Wyoming, all states rely on the severance tax for more than half of total state-local revenue from oil production. Moreover the property tax in Wyoming is actually an ad valorem levy on the value of annual production rather than the value of mineral rights. Thus it is indistinguishable from a severance tax. With this interpretation California's revenue structure is clearly atypical. Whereas all other states raise 60 to 100 percent of total (1980–81) tax revenue from severance taxes, California's severance tax is negligible.

Only two of the six states, California and Texas, base property taxes on the assessed value of proved reserves. In both states property tax receipts are a major source of revenue, accounting for 43 percent and 36 percent of (1980–81) statewide collections in California and Texas. Interpreting Wyoming's property levy to be a severance tax, none of the other major oil-producing states collects as much as 5 percent of total tax revenue from levies on property values.

Only California and Alaska rely on the corporate income tax for

TABLE 2.3

State and Local Government Tax Receipts from Crude Oil Production: Fiscal
Year 1979–1980
(millions of dollars)

Taxes	Alaska	Calif.	La.	Okla.	Tex.	Wyo.
Severance tax	497	—[a]	335	226	786	46
Property tax	68	99	17	0	342	118
Corporate income-franchise tax	381	278	68	34	26	—[a]
Total	947	377	421	260	1,154	164

SOURCES: See appendix B.

[a] Less than 0.5.

TABLE 2.4

State and Local Government Tax Receipts from Crude Oil Production: Fiscal
Year 1980–1981
(millions of dollars)

Taxes	Alaska	Calif.	La.	Okla.	Tex.	Wyo.
Severance tax	1,151	—[a]	639	320	1,291	131
Property tax	84	175	20	0	771	181
Corporate income-franchise tax	651	233	96	50	33	—[a]
Total	1,886	408	755	370	2,095	312

SOURCES: See appendix B.

[a] Less than 0.5.

substantial shares of overall petroleum-related tax revenue, and in California the corporate income tax accounts for almost 60 percent of (1980–81) state tax collections. Alaska is second in its reliance on the corporate income tax, with a 35 percent share in 1980–81, and Alaska's high corporate tax receipts during these years were a temporary phenomenon, attributable to a policy of "separate accounting" in force during 1979–81. The corporate tax rate applied in Alaska has changed twice since 1978. In 1978 the state of Alaska adopted legislation requiring separate accounting methods for petroleum companies; these firms were also subject to a special corporate income tax schedule with a maximum marginal tax rate of 11 percent. Nonpetroleum corporations in the state faced maximum marginal rates of 9.4 percent. This two-part tax structure remained in effect until 1981, when the legislature abolished separate accounting for oil companies and reduced the tax rate they faced to the 9.4 percent paid by all other corporations. The magnitude of this transitory effect is evident from the fact that the Alaska Department of Revenue (1985) reported total corporate income tax collections from the petroleum industry at $167 million in 1984–85, down from the 1981 level of $895 million.

The following general conclusions follow from the preceding comments on the structure of petroleum taxation in major producing states. Most states rely on the severance tax for the majority of oil-related tax revenues. California is the sole exception to this rule. Only two states, California and Texas, obtain significant shares of petroleum tax revenues from true property taxes, and in both states the yield is substantial, accounting for about one-third of statewide revenue. Corporate income taxes represent a relatively unimportant revenue source in all states except Alaska and California. In the latter, reliance on corporate income taxation is particularly heavy, since about 60 percent of oil-related revenue is attributable to this tax instrument.

To permit interstate comparisons of average tax rates, expressed as percentages of taxable production, table 2.5 presents data on the value of

TABLE 2.5
Value of Crude Oil Production by State and Tax Status
(millions of dollars)

	1979–1980		1980–1981	
	Total	Taxable[a]	Total	Taxable[a]
Alaska	$7,369.690	$7,335.818	$11,669.396	$11,604.749
California	6,270.764	5,037.529	9,192.043	7,291.493
Louisiana	7,316.097	3,280.182	12,742.647	5,692.901
Oklahoma	3,159.802	3,149.935	5,028.224	5,014.184
Texas	16,951.878	16,815.474	28,663.469	28,297.146
Wyoming	1,942.124	1,789.973	3,312.456	3,059.740

SOURCES: American Petroleum Institute, *Basic Petroleum Data Book* (Washington, D.C., 1982).

Independent Petroleum Association of America, *The Oil Producing Industry in Your State* (Washington, D.C., 1982).

NOTE: Calendar years averaged to obtain values on a fiscal year basis.

[a] Excluding production from federal properties.

crude oil produced in each state. Figures labeled "nonfederal" exclude production from federal O.C.S. leases, the federal royalty interest in oil produced on onshore federal leases and Indian lands, plus production on Naval Petroleum Reserves (Elk Hills N.P.R.). Nonfederal production is used as our base for comparisons of effective tax rates in various states.

Neither property nor severance taxes are levied on federal production, and of course no state royalties are received from it. Production on federal property does, however, generate some taxable corporate income. As noted in appendix *A*, a portion of income from O.C.S. production is taxable under unitary tax arrangements, and one might adjust the base for the corporate income tax alone to reflect this. In the case of California the required adjustment would be insignificant (about 1.1 percent); in Louisiana the result would be an approximate 41-percent increase in the corporate income tax base and a consequent reduction in the reported tax rate. In California two additional offsetting adjustments would be necessary, an increase in the base to reflect taxable income earned by the contractor who operates the Elk Hills N.P.R. for the U.S. Department of Defense, and a decrease to reflect the effective tax exemption that results from the profit share system on state tidelands. To avoid the confusion that would arise from applying different bases to different revenue sources, the comparisons presented below all employ the nonfederal figures in table 2.5. It should be kept in mind, however, that this will impart an upward bias to corporate income tax rates in Louisiana.

The average effective tax rates presented in table 2.6 were adjusted to reflect state income tax payments for royalty oil income. Since state financial reports do not detail state income taxes paid on royalty oil income, it was necessary to estimate these revenues. The amounts of royalties paid to private parties in each state will depend on the total value of private oil production, on the fraction of that production from properties owned in fee by individuals rather than by producing corporations, and on each state's individual income tax rate schedule. Estimates of the value of oil produced from privately owned properties are discussed in appendix *A*. Data on the ownership of properties by individuals rather than producing companies are unavailable. However, the practice of company ownership appears to be more common in California than elsewhere. To reflect this we have assumed that company ownership applies to 25 percent of nongovernmental production in California but is nonexistent in other states. Given the resulting estimates of production from private, individually owned properties, income tax revenues were computed by assuming that royalty payments were uniformly one-eighth of wellhead value, that royalty recipients were eligible for state percent-

TABLE 2.6
State and Local Government Taxes with Comparisons to Gross Sales Value
(millions of dollars)

	1979–1980			1980–1981		
	Value of Taxable Production	State/Local Taxes		Value of Taxable Production	State/Local Taxes	
		Total[a]	% Value[a]		Total[a]	% Value[a]
Alaska	7,336	947	12.9	11,605	1,886	16.2
California	5,038	412	8.2	7,291	459	6.3
Louisiana	3,280	432	13.1	5,693	774	13.6
Oklahoma	3,150	278	8.9	5,014	399	8.0
Texas	16,815	1,154	6.9	28,297	2,095	7.4
Wyoming	1,799	164	9.1	3,060	312	10.2
Total	37,418	3,387	9.1	60,960	5,925	9.7
Total, lower 48	30,082	2,440	8.1	49,355	4,039	8.2

[a] Includes estimated state income tax payments on royalty income.

age depletion allowances, and that the resulting income was taxed at the state's maximum marginal rate. (Computations are shown in appendix A.) The resulting estimates of income tax payments on royalties amounted to 0.70 percent of nonfederal production in California, 0.34 percent in Louisiana, and 0.57 percent in Oklahoma in fiscal year 1980–81. (State personal income taxes are not levied in Alaska, Texas, and Wyoming.)

Table 2.6 presents state and local government tax receipts, including estimated taxes on royalty income, for all six comparison states. Also reported are average effective tax rates, computed as state and local government receipts as a percentage of the wellhead value of nonfederal production. The average tax rates shown in the third and sixth columns of figures in this table provide an indication of the wide range of tax burdens found in major oil-producing states. In 1980–81 state tax revenue per dollar of production in Alaska was over twice as high as in Oklahoma or Texas and over 2.5 times as high as in California. In terms of 1980–81 collections Alaska is truly an aberration. Moreover the general system of public finances in Alaska is unlike that found in any other state. In 1980, for example, per capita receipts from all taxes in Alaska were almost $4,200 (Advisory Commission on Intergovernmental Relations 1981). Not only was this the highest in the United States, it was almost three times as high as the next highest state (New York) and over four times the national average.

It is interesting to note that effective average tax rates on petroleum are relatively high in Alaska and Wyoming, states where the economic base is not highly diversified and the level of industrial development is comparatively low. This lack of economic integration may dictate a heavy dependence on petroleum production for public sector income. In the more highly developed economies of California and Texas, on the other hand, petroleum related taxes are lower.

Comparing figures in table 2.6 to those derived from the representative firm approach indicates the magnitude of potential error with the latter approach. Consider the averages of corporations *A* and *B* estimates from table 2.1 as compared to the midpoints of the two fiscal year tax rates in table 2.6. The representative firm approach resulted in overestimating tax burdens in all states except California, where the effective average tax rate was underestimated. The degree of error is particularly severe in Oklahoma and California, where tax burdens were, respectively, overestimated and underestimated by about one-third.

2.5 State Government Royalty Income from Crude Oil Production: Interstate Comparisons

When considering the share of a state's mineral wealth made available to state and local governments, it is appropriate to examine both taxes and royalties, if only for the simple reason that a dollar raised from either source is equally valuable to the state. Furthermore, there is a necessary interrelationship between tax and royalty income that requires the two to be examined together. This relationship arises because the government's royalty interest is not subject to property or severance taxes; moreover, state royalty payments are deductible from corporate income taxes, and the state's royalty receipts generate no income tax revenue. If state royalties increase, tax revenues must necessarily decline. Consider, for example, California, where state royalties are particularly important. If California's royalty income were lower than it now is, tax receipts would necessarily be higher. To make this point in a slightly different manner, one might view the state of California as a large tax-exempt oil producer. In 1980–81 its operating interest amounted to production worth $476 million when valued at the wellhead. In these terms the state was the sixth largest producer in California but paid no tax.

Interstate comparisons of government royalty income are presented in table 2.7. The percentage figures show royalties expressed as a percent of the overall value of nonfederal production in each state. Thus these figures compare royalty receipts to that portion of each state's resource

TABLE 2.7

Government Receipts from Crude Oil Royalties: Interstate Comparisons
(millions of dollars)

	1979–1980		1980–1981	
	Royalty Revenue	% Statewide Value of Production[a]	Royalty Revenue	% Statewide Value of Production[a]
Alaska[b]	982	13.4	1,590	13.7
California	288	5.7	475	6.5
Louisiana[b]	167	5.1	287	5.0
Oklahoma	9	0.3	13	0.3
Texas[b]	205	1.2	348	1.2
Wyoming	15	0.8	25	0.8
Total	1,666	4.5	2,738	4.5
Total, lower 48	684	2.3	1,148	2.3

[a] Percent of the value of nonfederal production; see table 2.6 for production figures.

[b] Includes lease rentals and bonus payments; see appendix A.

base that has not been preempted, for revenue purposes, by the federal government. Judging from either total royalty revenues or percentages, it is evident that Alaska is again an outlier. State royalties in Alaska substantially exceed those in all other major oil-producing states combined. Moreover, royalty receipts as a share of nonfederal production are about six times as high in Alaska as in the lower 48.

California and Louisiana rank second and third in terms of total state royalty revenues and in terms of royalty receipts as a fraction of nonfederal production. State royalties are much smaller in the remaining comparison states and represent relatively minor claims on the value of nonfederal oil production.

2.6 Interstate Cost Comparisons

To this point the analysis in this chapter has compared government revenues in various states to the gross revenue derived from oil production. Comparisons to gross revenue alone do not, however, accurately portray the burden that a given revenue system imposes on resource owners and producers. The answers to such questions would be more accurately reflected in comparisons of public revenue to the net income derived from oil production. It is after all the effect of taxes and royalties on net income to the operator that is of primary concern in the decision of whether or not to explore and develop a new region, to incur the expense required to implement secondary and tertiary recovery projects, or to shut in a producing well. Likewise the net income associated with producing petroleum resources in a given state is a much better indicator of the economic value of those resources than is gross income. Accordingly, questions regarding the equity of the public's share in those resources would be much more accurately addressed by comparing government revenue to net income.

Available data do not permit direct computation of net income from oil

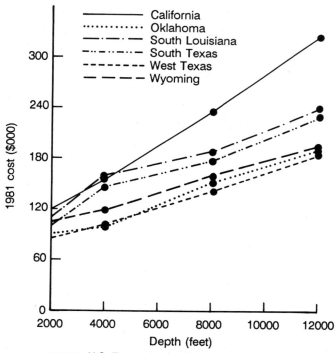

FIGURE 2.1
1981 Annual Operating Costs for Primary Oil Production,
by Depth, in Comparison Regions
(costs for 10 producing wells)

SOURCE: U.S. Energy Information Administration (1982).

FIGURE 2.2
1981 Lease Equipment Costs for Primary Oil Production,
by Depth, in Comparison Regions
(costs for 10 producing wells)

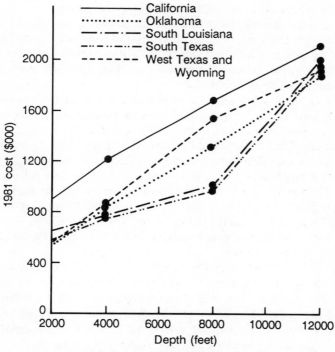

SOURCE: U.S. Energy Information Administration (1982).

production. There are, however, several regularly published indicators that can shed light on oil production costs in various states. The information portrayed in figures 2.1 and 2.2 refers, respectively, to monthly operating outlays and costs of lease equipment required for primary oil production. (Unfortunately production costs and lease equipment expenditure estimates are not available for Alaska.) The data represented in figure 2.1 indicate that operating costs in each depth range tend to be higher in the western United States (primarily California) and lower in west Texas and Oklahoma than in other regions. Higher production costs in California are attributable to several factors. Labor costs tend to be relatively high and safety and environmental regulations relatively stringent in California. Moreover, the low-gravity crudes extracted in California typically require heavier pumping equipment and larger expenditures of energy for lifting oil to the surface than in other states.

Figure 2.2 indicates relatively large lease equipment outlays in California and relatively low outlays in south Texas and south Louisiana. Again, the higher figures in California are likely attributable to the predominance of very heavy crude oil produced in that state and partially to environmental and safety regulations.

The data in figures 2.1 and 2.2 provide only partial information on production costs and the net income from oil production in major producing states. To obtain a more complete picture it would be necessary to modify and supplement these data in several respects. For example, cost estimates for various depth ranges should be adjusted to account for the actual producing depths in various states. It is generally agreed that producing depths in California (onshore) are less than in most other producing regions of the United States, and this would tend to reduce the apparently high cost of California operations. Data available from regularly published sources are not sufficient, however, to take this factor into account in a precise way.

Because the cost data shown in figures 2.1 and 2.2 are on a per well basis, they do not account for differences in the productivity of wells in various states. Table 2.8 presents data on production per well, price per barrel, and the gross value of output per well in various states. (Alaska was excluded due to the lack of operating cost and lease equipment outlays for that state.) Because average depth ranges for production in various states are unknown, it is still not possible to form estimated production and lease equipment costs per barrel. However, the vast differences in gross revenue per well, and the broad similarity in costs per well shown in figures 2.1 and 2.2 permit inferences regarding the general pattern of costs that would emerge from such computations.

Due to a predominance of stripper wells, Oklahoma has by far the lowest value of output per well of major producing states in the lower 48. This strongly suggests that operating and lease equipment outlays, per barrel, are higher in Oklahoma than elsewhere. Relatively low productivity in Texas and low field prices in California make these two states the second and third lowest in value of output per well, while Louisiana and Wyoming exhibit the two highest productivities per well.

Overall the preceding comparisons indicate that operating costs per

TABLE 2.8
Before-Tax Revenues per Well: Interstate Comparisons for 1981

	Production per Well (bbl./day)[a]	Crude Oil Price ($/bbl.)[a]	Gross Revenue per Well per Day
California	19.7	26.72	536
Louisiana	20.1	35.62	716
Oklahoma	5.1	35.64	185
Texas	14.4	35.34	509
Wyoming	31.1	32.49	1,010

[a] Independent Petroleum Association of America (1982).

dollar of gross revenue are relatively high in Oklahoma and relatively low in Louisiana and Wyoming, with California and Texas occupying a middle ground. Of course differences in exploration and development outlays could upset these rankings. To bring these cost factors into these comparisons would require converting such investment outlays into per barrel costs, adjusting for differential drilling costs, depths, reservoir sizes, dry-hole probabilities, and other factors, an exercise that is beyond the scope of the present analysis. Nevertheless it is interesting that the rough comparisons in table 2.8 are closely correlated with the estimated average tax rates in table 2.6. That is, effective state and local tax rates tend to be high in states where net operating revenues per well are relatively high. This suggests that taxes expressed as a fraction of net operating revenues would be more uniform than the pattern in table 2.6 indicates.

An additional cost consideration arises from the fact that the cost estimates in figures 2.1 and 2.2 refer only to primary production. Secondary and enhanced recovery are of course more costly and are becoming increasingly important in the United States. This is a particularly salient consideration in California, where a prevalence of heavy crude oils has promoted the widespread use of enhanced recovery by steam injection.

As tables 2.9 and 2.10 demonstrate, California is a leader in the use of steam injection and in situ combustion techniques. The state of California, which produces only about 11 percent of all crude oil extracted in the United States, accounts for about 78 percent of all enhanced oil recovery (E.O.R.) production. Not only does California dominate in this area, but production by E.O.R. methods accounts for a large and growing

TABLE 2.9
Enhanced Oil Recovery Projects in the United States

| | Active Projects | | | | | Planned |
	Steam	In Situ Combination	Gas Injection	Chemical Flood	Total	All Types
California	106	7	2	5	120	28
Louisiana	2	4	12	4	22	3
Oklahoma	3	0	3	14	20	8
Texas	5	5	20	22	52	1
Wyoming	2	0	3	16	21	0
Other	0	5	10	24	39	11
Total	118	21	50	85	274	51

SOURCE: *Oil and Gas Journal*, 5 April 1982.

TABLE 2.10
Enhanced Oil Production in the United States

Type	Production (bbl./day)	%	Production in Calif.	% of U.S.
Steam	288,396	77	285,042	99
In situ combustion	10,228	3	4,873	48
Chemical floods	4,409	1	520	12
Gas injection	71,915	19	360	0.5
Total	374,948	100	290,795	78

SOURCE: *Oil and Gas Journal*, 5 April 1982.

fraction of total production in the state. The 290,795 bbl./day attributed to enhanced recovery in table 2.10 amounts to about 28 percent of total California production during 1980.

The data on costs of production for various E.O.R. methods that would permit one to translate these observations into more refined cost estimates, and hence net income comparisons among states, are not presently available. It seems clear, however, that the prevalence of enhanced recovery in California significantly depresses net operating revenue per barrel. Thus comparisons based on net rather than gross income would probably raise California in the interstate ranking of tax burdens.

2.7 Conclusions

The preceding comparisons indicate that the tax burden on crude oil produced in California, when expressed as a fraction of the gross value of production, is at the low end of the range for major oil-producing states in the United States but is not markedly different from burdens found in Texas and Oklahoma. When expressed in this fashion, state and local tax burdens appear much higher in Wyoming, Louisiana, and Alaska. State royalties, however, represent relatively large fractions of the gross value of nonfederal production in Alaska, Louisiana, and California. When tax and royalty revenues are combined, total state and local revenue, expressed as a fraction of the gross value of production, are by far highest in Alaska, followed by Louisiana and California in that order. On this combined basis the state and local share of gross oil income appears lowest in Oklahoma and Texas.

Available data on production and investment outlays for oil production do not permit the state and local government "take" to be expressed as a fraction of the net value of resources. However, indications that productivities per well are highest in Alaska and Louisiana and lowest in Oklahoma, Texas, and California, indicate that taking costs into account would tend to produce a more uniform pattern of tax burdens and tax plus royalty burdens across states.

Perhaps the most dramatic difference in tax practices among the states concerns variations in tax structures and the extent to which particular tax instruments are relied on to raise revenues. As noted earlier, most states place heaviest emphasis on severance taxes, which represent levies against gross revenue. California, however, relies almost exclusively on property and corporate income taxes. The corporate income tax of course allows deductions for both operating costs and investment outlays. The property tax, as levied in California (and Texas), is in principle levied against the net value of mineral rights. Implicitly, then, operating costs are deducted when computing the property tax base. (Recall that the Wyoming property tax is effectively a tax on production.)

These broad differences in tax structure are displayed in more detail in table 2.11. If one made a similar distinction between state royalty collection practice, the sharp difference between California and other states would be even more striking. The most productive state leases in California were originally leased under a profit-sharing arrangement. Within this system state royalty collections are computed as a fraction of net lease revenue after deductions for costs. To the extent that accounting practices are similar, these royalty collections are comparable to receipts

TABLE 2.11
Comparison of Tax Structures in Various States
(1980–81)

	Taxes on Gross Revenue (%)[a]	Taxes on Net Revenue (%)[b]	Total (%)
Alaska	61	39	100
California	10	90	100
Louisiana	86	14	100
Oklahoma	87	13	100
Texas	62	38	100
Wyoming	100	0	100

SOURCE: See table 2.4.

[a] Severance taxes, individual income taxes on royalty income, and the property tax in Wyoming.

[b] Corporate income taxes and true property taxes.

from a corporate income tax on state levels. Other producing states, however, compute royalties as a fraction of the gross value of production (plus bonus payments for lease acquisition).

While all taxes depress incentives to invest in and produce marginal resources, the magnitude of the effect can vary sharply with the type of tax levied. According to the National Research Council, severance taxes have a more serious impact on resource development than do equal yielding taxes on net income:

> [T]he severity of the investment dampening effects of taxation on new investment in the energy resources sector are not the same for all taxes that yield the same revenue or for all types of energy resources. . . . Income taxes are less depressing on exploration expenses than are production [severance] taxes. This is because under the usual form of income tax, exploration expenditures enter into the determination of taxable income. . . . Production taxes in their usual form, however, do not allow any recognition of exploration costs. . . . Production taxes also impinge more heavily on investments in the development of already discovered deposits. . . . Production taxes involve more significant effects on development decisions than do income taxes as they tend to augment the risks perceived by the firm. When there are production taxes, additional output made possible by investment in new development will give rise to tax liability whether the operation is profitable or not. In the case of an income tax, tax liability is incurred only in the presence of profits. . . . To the extent that further reserves of energy resources lie in deeper and less accessi-

ble formations, these reserves can be tapped only with substantial development outlays. Indeed, such outlays are likely to be more substantial as a proportion of total investment in the energy industry in the future than in the past. Reliance on production taxes will clearly be less favorable to such development intensive investment than would be income taxation, and therefore less favorable to expansion of national energy supplies (U.S. Energy Information Administration 1980, pp. 55–57).

The prevalence of federal and state tax exemptions or rate reductions for stripper oil, heavy oil, newly discovered petroleum, and so forth, reflects precisely such concerns. Yet these exemptions can at best provide only rough approximations to the ideal, a full recognition of cost differences in the definition of the tax base.

As presently constituted all three of California's major revenue instruments, property taxes, corporate income taxes, and royalties, make allowances for costs. Moreover, California is unique in this respect among major producing states. Whether by accident or design, California now has a revenue structure that is relatively efficient, particularly in light of the economically marginal nature of much of the state's crude oil resource.

Appendix A

Computing Tax Burdens: Sources of Data and Methods of Estimation

This appendix is arranged in four parts. The first describes sources of data for crude oil severance tax receipts, state and local government property tax payments by the oil production industry, and royalty payments to state governments. The second section outlines the method used to estimate corporate income tax payments by oil producers. Of the six states examined, only Alaska, California, and Texas report corporate income or franchise tax payments by industry. Information outlined in section A.1 enabled estimation of corporate franchise tax payments in the state of Wyoming. Corporate income taxes in the remaining two comparison states, Oklahoma and Louisiana, are not reported on an industry-by-industry basis and thus were estimated. The third section presents sources of data for production by state, and for production on federally owned property. Data on the gross wellhead value of (1) total production and (2) taxable production (which excludes production from federal properties) are also presented. The fourth section presents data and assumptions used to estimate state income tax payments on royalty income.

A.1 Data Sources and Estimation Methods for Individual States

For virtually all revenue sources except corporate income and franchise taxes, it was possible to obtain actual revenues from various published government reports. In some instances data were reported only for crude oil and natural gas combined. In such cases receipts were apportioned to the two resources on the basis of the gross value of taxable production in the particular state. Where data were reported by calendar year, figures for successive years were averaged to approximate fiscal year receipts.

Appendix A

A.1.1   Alaska

Severance tax receipts for crude oil and natural gas are reported by the Alaska Department of Revenue (1983) for both fiscal years examined. The figures reported in table A.1 are for crude oil only and include relatively minor receipts from the crude oil conservation tax.

Property taxes on oil and gas property are levied by both the state of Alaska and local governments. The top three rows of figures in table A.2 show state and local receipts for property taxes levied on exploration equipment, production equipment (including gathering lines), and pipelines (including terminals, pumping stations, and so forth). To maintain comparability with property tax figures reported for other states, the portion of tax payments attributable to pipelines was excluded from total property tax payments. As shown in table A.2, this was accomplished by assuming that the share of tax payments attributable to pipelines was equal to the corresponding assessed value share. The tax estimates reported in the last row of table A.2 were then adjusted to exclude payments attributed to natural gas properties, and the resulting figures appear in the second row of table A.1.

The Alaska Department of Revenue (1983) reports data on corporate income taxes paid by oil and gas producers. The same agency was able to provide estimates of the percentage of tax receipts attributable to production as opposed to pipelines and other activities. For fiscal years 1979–80 and 1980–81, respectively, 71 percent and 77 percent of corpo-

TABLE A.1
State and Local Government Receipts from Crude Oil Production: Alaska
(millions of dollars)

	1979–1980	1980–1981
Severance tax	$497	$1,151
Property tax	68	84
Corporate income tax	381	651
Total tax	947	1,886
State royalty receipts	982	1,590
Total tax plus royalties	$1,929	$3,476

SOURCES: Alaska Department of Revenue, *Revenue Sources FY 1982–83* (Juneau, Alaska: January 1983). Alaska State Assessor's Office, *Alaska Taxable Property*, various years (Juneau, Alaska). Includes rental payments and average bonus receipts; see text for further explanation.

NOTE: See text for discussion of adjustments and additional data sources.

TABLE A.2

Property Taxes on Oil and Gas: Alaska

(millions of dollars)

	1979–1980	1980–1981
Property tax levies		
State	$165.725	$140.976
Local	66.648	126.705
Total	$232.373	$267.705
Percent of statewide assessed value in pipeline[a]	70%	68%
Property tax attributable to exploration and production equipment	$69.712	$85.658

SOURCES: Alaska State Assessor's Office, *Alaska Taxable Revenue,* various years (Juneau, Alaska).

Personal communication, Mr. Jerry Heier, Alaska Department of Revenue, Petroleum Revenue Division, 2 May 1983.

[a] Estimate provided by Alaska Department of Revenue.

rate tax receipts from this industry were attributable to crude oil and natural gas production. As before, the market value share of crude oil in total hydrocarbons was used to prorate the resulting income tax payment between crude oil and natural gas.

Royalty and rental payments to the state of Alaska's general fund are reported by the Alaska Department of Revenue (1983). The portion attributable to crude oil was estimated using the allocation formula previously explained. The royalty payment figures in table A.1 also include estimates of the values of nonrecurring bonus payments for new state leases. The bonus payments were computed from data on actual lease sales for the period from 1971 to 1981. Bonus receipts in each year were expressed in equivalent barrels of oil by dividing dollar payments by the price of Alaskan crude oil in each year. The resulting bonus payments measured in barrels of oil were then averaged. Average bonus payments for the two fiscal years reported in table A.1 were obtained by multiplying the 10-year average quantity bonus by the price of crude oil during these two fiscal years. The result was an average bonus payment of $44.380 million in fiscal year 1979–80 and $66.569 million in fiscal year 1980–81. Since 1977 a portion (25 percent prior to 1982) of all royalty, rent, and bonus payments collected by the state of Alaska has been placed in the Alaska Permanent Fund. Because the figures just reported reflect only General Fund receipts, it was necessary to increase them (by one-third) to incorporate Permanent Fund collections. Thus the royalty payments shown in table A.1 include receipts by both funds.

Some of the revenue collected by Alaska in 1980 includes receipts from the sale of leases in the Beaufort Sea. The federal government has disputed Alaska's jurisdiction in this sale, and a legal decision on the matter is now pending. Depending on the outcome of this litigation, crude oil revenues attributable to the state of Alaska in 1980 are subject to possible revisions.

A.1.2 California

Severance tax receipts from oil and gas production in California are reported by the California state comptroller. The 1980 shares of gross output value were used to attribute these receipts to crude oil and natural gas.

The assessed value of mineral rights and associated improvements for production plus the total property tax levied by local governments on these values are reported each fiscal year by the California State Board of Equalization. The property tax figures shown in table A.3 were taken from this source.

California reports corporate income tax receipts by industry; thus actual data were obtained on corporate income and franchise tax payments by firms engaged in oil and gas production. These data from the California Franchise Tax Board relate to the 1980 and 1981 "income

TABLE A.3
State and Local Government Receipts from Crude Oil Production: California
(millions of dollars)

	1979–1980	1980–1981
Severance tax[a]	$0.283	$0.467
Property tax[b]	98.858	174.936
Corp. income-franchise tax[c]	277.845	232.854
Total tax	376.986	408.257
State royalty receipts[a]	288.185	475.403
Total tax plus royalties	$665.171	$883.660

[a] California State Comptroller, *Annual Report 1980–81* (Sacramento, Calif.).

[b] California State Board of Equalization, *County Oil and Gas Survey, 1981* (Sacramento, Calif.).

[c] California Franchise Tax Board, *Annual Report,* various issues (Sacramento, Calif.).

years," which correspond to fiscal years 1979–80 and 1980–81. Tax payments for oil and gas production were allocated between oil and gas on the basis of shares of the gross value of output for these two resources in 1980 (93.9-percent oil and 6.1-percent natural gas; Independent Petroleum Association of America 1981).

Royalty and rental payments to the state of California for production of crude oil and natural gas are reported by the California state comptroller. To split these total receipts between oil and gas, data were taken from the U.S. Geologic Survey (1982) on the offshore production of crude oil and natural gas on state tidelands. Average price data for oil and gas were obtained from the American Petroleum Institute (1982). Using both sources the fraction of total state tidelands production attributable to crude oil was placed at 95.6 percent in 1980. This share was then used to estimate royalty revenue attributable to crude oil alone.

A.1.3 Louisiana

The Louisiana Department of Revenue and Taxation reports severance tax receipts for oil and gas separately. The method used to estimate corporate income and franchise tax payments by Louisiana oil and natural gas producers is explained in a separate section of this appendix.

Property taxes in Louisiana are levied at the parish level, and tax rates vary widely among parishes. By law taxes may not be assessed on the value of mineral rights or production; only the value of production equipment is taxable. The Louisiana State Tax Commission reports assessed valuation of various types of production equipment as well as overall assessed value and property tax payments in each parish.

In order to estimate property tax payments by oil and gas producers, it was first necessary to obtain an effective overall property tax rate to be applied to oil and gas equipment. Data on total assessed valuations and total property tax receipts were collected for a sample of parishes that account for most of the statewide assessed valuation of oil and gas wells. The parishes included were Cameron, Lafourche, Plaquemines, St. Mary, Terrebonne, and Vermilion. Within this sample of parishes total valuations and property tax collections for oil and gas wells were $143.915 million and $9.002 million, respectively, in fiscal 1981. These figures imply an effective property tax rate of 6.26 percent for parishes that account for the bulk of the state's oil and gas well assessed value.

To obtain an estimate of statewide property tax payments for oil and gas equipment, the preceding tax rate was applied to statewide figures for individual items of oil and gas production equipment. The relevant data, for fiscal years 1979–80 and 1980–81, are shown in table A.4. When applied to the tax rate just shown, total property tax receipts for oil and gas are estimated at $28.147 and $30.133 million for the two years. The share of these receipts attributable to crude oil equipment was set equal to the share of oil and gas wells in the state that are primarily oil producers (62 percent in 1980 and 66 percent in 1981). The resulting property tax liability for crude oil production is reported in table A.5.

State rent and royalty receipts for crude oil and natural gas are reported by the Louisiana Division of Administration. Because oil and gas

TABLE A.4
Statewide Assessed Valuation of Oil and Gas Production Equipment: Louisiana
(millions of dollars)

	1979–1980	1980–1981
Oil and gas wells	$242.941	$246.512
Equipment		
Oil tanks	12.372	13.476
Pipes, gathering lines	81.383	91.539
Drilling rigs	42.065	52.649
Oil and gas lease equipment	71.050	77.179
Total	$449.629	$481.346

SOURCE: Louisiana State Tax Commission, *Annual Report* (Baton Rouge, La.).

TABLE A.5
State and Local Government Receipts from Crude Oil Production: Louisiana
(millions of dollars)

	1979–1980	1980–1981
Severance tax[a]	$335.017	$639.024
Property tax[b]	17.350	20.032
Corp. income-franchise tax[c]	68.369	95.920
Total tax	420.736	754.976
State royalty receipts[d]	166.718	286.998
Total tax plus royalties	$587.536	$1,041.974

[a] Louisiana Department of Revenue and Taxation, *Annual Report* (Baton Rouge, La.).

[b] Louisiana Tax Commission, *Annual Report* (Baton Rouge, La.); see text.

[c] See text.

[d] Louisiana Division of Administration, *State of Louisiana, Annual Financial Report* (Baton Rouge, La.); see text.

figures are not reported separately, it was necessary to estimate the share attributable to crude oil alone. To do so, information from the U.S. Geologic Survey (1982) was used to compute the value of oil and gas produced on Louisiana state offshore lands. The share of this value represented by crude oil was used to estimate state crude oil royalty receipts.

The figures reported in table A.5 include estimates of the values of nonrecurring bonus payments for new state leases. These payments were computed from data on actual bonus payments over the period from 1973–74 through 1981–82 using the method outlined in the discussion of Alaskan royalty and bonus receipts. The results were estimated average crude oil bonus payments of $54.870 million in 1979–80 and $90.603 million in 1980–81.

A.1.4 Oklahoma

Total severance tax receipts for oil and gas produced in the state of Oklahoma are reported by the Oklahoma Tax Commission. Officials of the tax commission provided separate figures for crude oil and natural gas in personal communication. Because only totals for oil and gas royalties were available, it was necessary to estimate the share attributable to crude oil. This share was assumed to equal the share of crude oil in oil and gas severance tax receipts. Although property taxes are levied by local jurisdictions in Oklahoma, these taxes do not apply to crude oil production, reserves, or production equipment.

The method used to estimate corporate franchise tax payments by crude oil producers is presented elsewhere in this appendix. State and local receipts from oil production in Oklahoma are shown in table A.6.

A.1.5 Texas

Severance tax payments are reported separately for oil and gas by the Texas state comptroller of public accounts. This agency also reports state receipts of oil and gas royalties for production on state land. The royalty receipt figures reported in table A.7 include estimated average annual bonus payments for crude oil extraction rights on state land. These averages were computed by applying the method used for Louisiana. Estimated crude oil bonus and rental payments for 1979–80 and 1980–81 were placed at $51.781 million and $106.717 million, respectively.

Local governments in Texas levy property taxes on crude oil and natural gas reserves in their jurisdictions. The base of the tax is the value

TABLE A.6
State and Local Government Receipts from Crude Oil Production: Oklahoma
(millions of dollars)

	1979–1980	1980–1981
Severance tax[a]	$226.278	$320.361
Property tax[b]	—	—
Corp. income-franchise tax[c]	34.420	49.831
Total tax	260.698	370.192
State royalty receipts[c]	8.766	12.867
Total tax plus royalties	$269.464	$383.059

[a] Oklahoma Tax Commission, *Annual Report for the Fiscal Year Ending June 30, 1981* (Oklahoma City, Okla.); and personal communication from Jill Rooker, Oklahoma Tax Commission, 6 December 1982.

[b] Does not apply.

[c] See text.

TABLE A.7
State and Local Government Receipts from Crude Oil Production: Texas
(millions of dollars)

	1979–1980	1980–1981
Severance tax[a]	$785.700	$1,290.900
Property tax[b]	342.027	770.693
Corp. income-franchise tax[c]	25.917	33.408
Total tax	1,153.644	2,095.001
State royalty receipts[a]	205.421	348.843
Total tax plus royalties	$1,359.065	$2,443.844

[a] Texas Comptroller of Public Accounts, *Annual Financial Report, 1981, State of Texas* (Austin, Tex.).

[b] Personal communication, Tim Wooten, coordinator for minerals, Texas State Property Tax Board, Austin, Tex. (see text), 14 December 1982.

[c] Personal communication, Stuart Greenfield, Revenue Estimating Division, Texas State Comptroller's Office, Austin, Tex., 14 December 1982.

of mineral rights as estimated by the discounted value of expected future income from production. Tax rates vary widely among the various local jurisdictions in the state. Estimates of total property tax receipts from all taxing jurisdictions (cities, counties, school districts, and special districts) were supplied by the Texas State Property Tax Board. Property tax receipts for all minerals, including oil, gas, and some nonfuel minerals, were estimated at $549 million in 1979–80 and $1,107 million in 1980–81. These receipts were apportioned between oil and gas on the basis of the market value of oil and gas production in these years (as reported by the American Petroleum Institute 1982).

Payments of corporate franchise taxes by oil producers were estimated from data supplied by personal communication from the Revenue Estimating Division of the Texas Comptroller's Office. This agency reported that firms engaged in crude oil and natural gas production (Standard Industrial Classification 13) were assessed franchise taxes of $41.6 million in 1979–80 and $48.0 million in 1980–81. Oil production taxes were apportioned in accordance with the share of crude oil in the value of statewide oil and gas production.

A.1.6. Wyoming

Severance tax payments for crude oil extraction are reported by the Minerals Division, Wyoming State Department of Economic Planning and Development. That agency reports severance tax liabilities on a calendar year basis for years in which production took place. (Taxes are actually paid during the following fiscal year.) These calendar year taxes were converted to a fiscal year basis by averaging over successive calendar years. It should be noted that the severance tax rate on crude oil was raised by 2 percent in January 1981. Partly for this reason, state projections estimate that crude oil severance tax receipts during fiscal 1982–83 would be over 40 percent higher than actual receipts during 1980–81.

Counties and other local governments in Wyoming levy what is termed a property tax (ad valorem) on crude oil and other minerals. In practice, however, the tax is levied on the value of the resource after extraction (rather than the value in situ), so that this source of revenue is indistinguishable from a locally applied severance tax. The items shown in table A.8 were taken from Wyoming Legislative Service Office (1982) and represent averages of calendar year figures.

Wyoming levies a very small corporate franchise tax (1980 receipts from all industries were $585,000). The tax is based on the value of assets located in the state. According to a publication by the Wyoming League of Women Voters (1981), the mineral industry accounts for about 33 percent of all state revenue, and crude oil accounted for 51 percent and 59 percent of this amount in 1980 and 1981, respectively. These figures were used to attribute 16.8 percent and 19.3 percent of corporate franchise taxes to crude oil in 1980 and 1981, respectively. Total franchise tax

TABLE A.8

State and Local Government Receipts from Crude Oil Production: Wyoming
(millions of dollars)

	1979–1980	1980–1981
Severance tax[a]	$45.719	$130.530
Property tax[b]	118.143	181.245
Corp. income-franchise tax[c]	0.098	0.157
Total tax	163.960	311.932
State royalty receipts[d]	14.638	25.336
Total tax plus royalties	$178.598	$337.268

[a]Wyoming State Department of Economic Planning and Development, *1981 Wyoming Minerals Yearbook* (Cheyenne, Wyo.: 1981).

[b]Wyoming Legislative Services Office, Management Council, *Taxes on Wyoming's Minerals: History and Projections* (Cheyenne, Wyo.: September 1982).

[c]Wyoming Department of Administration and Fiscal Control, *Wyoming Data Handbook, 1981* (Cheyenne, Wyo.: 1981).

[d]Personal communication, Helen Deniston, Wyoming State Auditor's Office, Cheyenne, Wyo., 10 December 1982.

receipts in Wyoming were obtained from the Wyoming Department of Administration and Fiscal Control (1981) and by personal communication from the Wyoming State Auditor's Office. Wyoming does not levy any tax on corporate income earned in the state.

State royalty payments for oil and gas for fiscal years 1980 and 1981 were obtained from the Wyoming State Auditor's Office. These were allocated to oil and gas on the basis of the assessed value of oil and natural gas produced in the state. (In 1980 and 1981 crude oil accounted for 74 percent and 80 percent, respectively, of total oil and gas valuation.)

A.2 Corporate Income and Franchise Tax

Corporate income and franchise taxes for Louisiana and Oklahoma were estimated with the procedure explained in this section. Throughout this section corporate income and franchise taxes are both considered as taxes on corporate income.

The estimation method employed is a technique for disaggregating data in a consistent way. The method is consistent in two respects: first, the procedure ensures that estimated corporate tax payments by individual industries in a given state will necessarily sum to the known total of corporate tax payments in that state. Second, our estimates of a given industry's taxable corporate income in each state are constructed so that if summed across all states in the United States, the total would necessarily coincide with the known level of taxable corporate income in the United States. As already noted, data on corporate income taxes paid in Alaska, California, and Texas by oil and gas producers are reported by the state; thus estimation was unnecessary. Despite this, estimates were formed for California using the technique outlined in the following paragraphs as a check to see how these estimates would compare to actual values.

To explain our estimation method in more detail, it is useful to introduce the following notation:

T_{is} is tax levied (in dollars) on corporate income in industry i in state s.
I_{is} is taxable corporate income in dollars for industry i in state s.
x_{is} is a measure of the activity level at which industry i operates in state s (explained more fully in the following paragraphs).

By definition, T_{is} may be expressed as

$$T_{is} = (T_{is}/I_{is})(I_{is}/x_{is})x_{is} \qquad (A.1)$$

Note that this expression is true regardless of the measure of activity chosen. In fact here is no need to use the same activity level measure for different industries. In the estimates presented here, the activity indicator chosen for all industries except oil and gas production is employment. For oil and gas the activity measure used is the value of sales.

To construct corporate tax estimates, I_{is}/x_{is} for each state is assumed to equal the national total for income per unit activity by industry. That is,

$$I_{is}/x_{is} = I_{in}/x_{in} \qquad (A.2)$$

where the subscript n refers to the national total. For nonpetroleum sectors U.S. taxable income per employee is specified to equal taxable income per employee in each state examined. For oil and gas production taxable income per unit sales in each state examined is specified to equal income per unit sales for the United States. In this fashion available data on taxable income and employment at the national level together with the data on employment by state and industry were used to estimate taxable income by state for nonoil and gas production sectors. As already noted, oil and gas production income was distributed among states on the basis of total sales rather than employment. (This seemed appropriate, since due to the large price changes that took place recently, income per employee has shifted dramatically.)

It is possible that some bias will result for individual states if, for example, profit per dollar of sales is abnormally high or low in a particular state. The same is clearly true for income per employee in nonextractive industries. But in the aggregate these errors must cancel out. This can be seen by writing out the estimate I'_{is} of income in industry i and state s,

$$I'_{is} = (I_{in}/x_{in})x_{is} \qquad\qquad (A.3)$$

If these estimates are summed across all states, the result is

$$\sum_s (I_{in}/x_{in})x_{is} = (I_{in}/x_{in})x_{in} = I_{in} \qquad\qquad (A.4)$$

In each sector, then, the sum of all state corporate income estimates must equal national corporate income.

The data needed to form these estimates of income by state and industry are presented in detail in Deacon et al. (1983). Before using these figures to estimate corporate tax liabilities by state and by industry, it was necessary to make two adjustments. First, our corporate income estimates include earnings from production on federal offshore leases. However, under the unitary income taxation procedures used in California and Louisiana, only a portion of this income is actually subject to the state income tax. With the unitary taxation formation, the fraction of a firm's total income that is taxable in a particular state is equal to the average of the firm's shares of total sales, employment, and assets that occur within that state. The following equation depicts the firm's taxable income in state s:

$$I_s = (1/3)(S_s + E_s + A_s)Y \qquad\qquad (A.5)$$

where

I_s is the income of the firm that is taxable in state s;
S_s is the fraction of the firm's sales that occurs in state s;
E_s is the fraction of the firm's employment that occurs in state s;
A_s is the fraction of the firm's assets that are located in state s; and
Y is the firm's total income.

TABLE A.9

Corporation Income and Franchise Taxes on Oil Production
(millions of dollars)

	1979–1980	1980–1981
California (estimated)	$269.327	$356.619
California (actual)	277.845	232.853
Louisiana	68.369	95.920
Oklahoma	34.420	49.831

SOURCE: See text.

In the case of production from federal O.C.S. leases, the employment and assets involved in production are, for tax purposes, located outside of the states' boundaries. Here E_s and A_s terms associated with production from O.C.S. properties are equal to zero. If the oil produced offshore is actually landed and sold in the adjoining state (for example, transported ashore by pipeline), then S_s in equation (A.5) would equal one for production from offshore fields. Thus the share of income from O.C.S. properties that is taxable in adjoining states would equal one-third. Accordingly, the estimates of taxable income in states that adjoin offshore properties (California and Louisiana) were adjusted to exempt two-thirds of income earned from federal O.C.S. properties from income and franchise taxes.

A second adjustment was required to reflect the fact that each of the states that levies a corporation income tax also allows percentage depletion deductions from taxable income. In California and Oklahoma the depletion allowances are 22 percent; in Louisiana the allowance is 38 percent. Because there are also limitations on the fraction of net income that may be "depleted," the effective depletion rates actually employed are typically less than those listed. To reflect these limitations the depletion rates were reduced by one-fourth. This adjustment is based purely on judgment, since there are no data on which to base a more accurate estimate. However, our detailed estimates indicate that interstate tax liability comparisons are not highly sensitive to this assumption.

With the preceding adjustments, taxable incomes for fiscal years 1979–80 and 1980–81 were computed for each state and industry. The shares of statewide corporate income attributable to crude oil were then multiplied by actual total corporate income tax payments in each state. The resulting estimates of corporate tax liabilities for crude oil production appear in table A.9. For comparison, actual corporate tax liabilities for California are shown as reported by the California Franchise Tax Board. The actual figures differ from estimates only slightly for 1979–80; in 1980–81, however, actual collections in California were about 35 percent below estimated levels. This suggests that the corporate income tax figures reported in the text for Louisiana and Oklahoma may be unrealistically high.

TABLE A.10
Tax Exempt Production from Federal Lands
(bbl.)[a]

	1979	1980	1981
Alaska			
Onshore	4,555,000	3,937,000	3,095,000
Offshore	—	—	—
Total	4,555,000	3,937,000	3,095,000
California			
Onshore	56,831,550	60,703,635	65,510,489
Offshore	10,961,076	10,198,885	16,406,976
Total	67,792,626	70,962,521	81,917,465
Louisiana			
Onshore	248,721	242,452	161,113
Offshore	271,008,916	256,668,082	255,875,717
Total	271,257,637	256,910,534	256,036,820
Oklahoma			
Onshore	503,252	445,307	446,128
Offshore	—	—	—
Total	503,252	445,307	446,128
Texas			
Onshore	17,300	22,848	18,070
Offshore	3,595,546	10,502,007	14,284,661
Total	3,612,846	10,524,855	14,302,731
Wyoming			
Onshore	9,982,375	9,426,750	9,341,112
Offshore	—	—	—
Total	9,982,375	9,426,750	9,341,112

SOURCES: U.S. Geologic Survey, *Outer Continental Shelf Statistics,* 1979–1981: (Washington, D.C.: U.S. Department of the Interior, 1982).

U.S. Minerals Management Service, *Royalties* (Washington, D.C.: U.S. Department of the Interior, 1982).

[a] Includes federal O.C.S. production, onshore production from Elk Hills Naval Petroleum Reserve, and federal royalty oil from onshore federal lease lands and Indian reservations.

TABLE A.11
Crude Oil Production
(MMBBL)

	1979		1980		1981	
	Total	Federal	Total	Federal	Total	Federal
Alaska	511.355	4.555	591.646	3.937	587.811	3.095
California	352.268	67.793	356.644	70.963	384.773	81.917
Louisiana	489.687	271.258	466.964	256.911	461.413	256.036
Oklahoma	143.642	0.503	151.960	0.445	165.426	0.446
Texas	1,018.094	3.613	975.239	10.524	1,026.359	14.302
Wyoming	131.890	9.982	129.309	9.427	118.844	9.341

SOURCES: American Petroleum Institute, *Basic Petroleum Data Book* (Washington, D.C.: 1982); see table A.12.

A.3   Quantity and Value of Taxable Crude Oil

In each of the six states compared, significant quantities of total oil production are from federal properties and are therefore tax exempt. Table A.10 shows quantities of tax exempt oil produced in each state for the years 1979 to 1981. All of the offshore production shown is from federal leases on the O.C.S. The onshore production figures include federal royalty interests from federal lease lands, royalty interests for production on Indian reservations, and production from the Elk Hills N.P.R. Total production is compared to production from federal properties in table A.11.

Table A.12 shows the value of total crude oil production in each state, derived by applying statewide average prices to total production in each state. The value of tax exempt federal production, computed in the same fashion, was subtracted from totals to obtain taxable values. Calendar year data for successive years were averaged to place them on a fiscal year basis.

A.4   Estimated State and Individual Income Taxes from Royalty Production

The royalty income tax comparisons presented in the text required estimation of the value of private oil production (excluding all state and federal government properties) in California, Louisiana, and Oklahoma in fiscal year 1980–81. Private royalties in all states were assumed to equal one-eighth of the gross value of production from private lands. Thus the first step in estimating income taxes from royalty receipts was to estimate the value of production from private lands.

For California production from state government leases was taken from the Conservation Committee of California Oil Producers (1982, pp. 24, 28); according to these figures state leases accounted for about 13.5 percent of all nonfederal production in California in 1980–81. The value of state production was estimated by applying this percentage to the value of nonfederal production. The resulting state production value is $986 million. (The estimate of state royalty production value will be somewhat low to the extent that the price of state royalty oil exceeds the statewide average. Given the magnitudes involved, however, this would not significantly affect royalty tax estimates.) Subtracting this sum from the value of nonfederal production yields an estimated private production value of $6,305 million for 1980–81.

Data on the value of production from state lands were unavailable for Oklahoma and Louisiana. Hence these production values were inferred from information on state royalty receipts. For Louisiana and Oklahoma it was assumed that the royalty rate on state government leases was one-

TABLE A.12
Value of Crude Oil Production by State and Tax Status
(millions of dollars)

Excluding Federal Taxes	1979–1980		1980–1981	
	Total	Taxable	Total	Taxable
Alaska	$7,369.690	$7,335.818	$11,669.396	$11,604.749
California	6,270.764	5,037.529	9,192.043	7,291.493
Louisiana	7,316.097	3,280.182	12,742.647	5,692.901
Oklahoma	3,159.802	3,149.935	5,028.224	5,014.184
Texas	16,951.878	16,815.474	28,663.469	28,297.146
Wyoming	1,942.124	1,789.973	3,312.456	3,059.740

SOURCES: American Petroleum Institute, *Basic Petroleum Data Book* (Washington, D.C.: 1982).

Independent Petroleum Association of America, *The Oil Producing Industry in Your State* (Washington, D.C.: 1982).

NOTE: Calendar years averaged to obtain values on a fiscal year basis.

TABLE A.13

Estimated State Income Tax Payments for Private Royalties: Fiscal Year 1980–1981

(millions of dollars)

	California	Louisiana	Oklahoma
Estimated private royalty payments[a]	591	516	614
Taxable royalties (net of depletion allowance)[b]	461	320	479
Tax at maximum marginal rate[c]	51	19	29
Tax as percent of value of nonfederal production	0.70%	0.34%	0.57%

NOTE: See text for further discussion and interpretation.

[a] Assumes that 25 percent of private production in California is from properties owned in fee by oil-producing corporations. No taxable private royalties would be earned on such production. It is assumed that there are no company-owned properties in Louisiana or Oklahoma.

[b] Assumes that royalty recipients would be eligible for percentage depletion allowances of 22 percent in California and Oklahoma and 38 percent in Louisiana.

[c] Assumes all taxable royalty income is taxed at 11 percent in California and 6 percent in Louisiana and Oklahoma.

eighth, the same as the assumed private rate. Thus state plus private royalties were computed by multiplying the value of nonfederal production in each state by one-eighth. From this total state royalty receipts (as reported by both states) were deducted to obtain estimated private royalty income.

Royalty income estimates plus the remaining steps required to obtain estimated income taxes from private royalties are presented in table A.13.

Chapter 3

Energy Tax Exporting:

A Dilemma for the Federal System

Federal systems have an economic advantage over centralized governments because local levels of government can produce the level and type
of local public goods that match the preferences of local residents. There
are also diseconomies of administration in government, so that productive efficiencies presumably result from decentralized systems.

However, federal systems also create many different kinds of inefficiency, particularly in taxation. Each state in a federal system has the
incentive to shift the burden of taxation to residents of other states and to
shift the benefits of government programs away from these out-of-state
residents. In what follows we will ignore the problems of benefit
spillovers across state lines and concentrate on taxes. As we will see, the
temptation to export taxes is particularly strong for taxes on immobile
energy resources.

The dangers of beggar-thy-neighbor state policies were clearly seen by
the founding fathers. Thus they prohibited state governmental policies
that inhibited interstate commerce. It might seem reasonable to interpret
this as simply prohibiting tax exportation. However, research discussed
later shows that all taxes are exported to a significant degree—especially
business taxes. Here is the dilemma. Federalism requires that states be
allowed to tax. Yet the power to tax implies, to some extent, the power to
export taxes. In what follows we will discuss the Supreme Court's recent
decision on *Commonwealth Edison* v. *Montana* (1981), where a divided
court undermined the constitutional restraints on state tax exporting.
Then we will survey the economic literature on tax exporting and apply
the lessons learned to a California oil severance tax.

3.1 The Law

In 1975 Montana raised its coal severance tax to 30 percent of the contract price. The tax was clearly intended to be paid mostly by out-of-state individuals. The high rate induced a group of Montana coal producers and electrical utilities from several states to challenge the tax in court. Further, these named plaintiffs were joined by the coal-consuming states of Iowa, Kansas, Michigan, Minnesota, New Jersey, Texas, and Wisconsin, which filed briefs as amici curiae at the appellate stage (Fish 1982, p. 1033).

The plaintiffs made two claims. First, the tax was contrary to the interstate commerce clause of the U.S. Constitution because it was tailored to and did, in fact, fall heavily on out-of-state users and because it was grossly out of proportion to the services provided by the state to the coal industry. Second, it violated the supremacy clause of the Constitution because it thwarted the expressed policy of the Congress to encourage the domestic use of coal. The Montana trial court granted summary judgment to the state and the State Supreme Court affirmed. The case was appealed directly to the U.S. Supreme Court. The appellants asked the court to reverse the summary judgment and remand the case for trial.

The court agreed with the appellants that the Montana coal severance tax "substantially affects interstate commerce" and therefore must be evaluated under a four-part test first enunciated in *Complete Auto Body v. Brady* (1977). This decision held that a state tax that affects interstate commerce is constitutional if it (1) applies to an activity having a substantial nexus with the taxing state, (2) is fairly apportioned (between the taxing state and other states), (3) does not discriminate against interstate commerce, and (4) bears a fair relationship to the services provided by the taxing state.

The entire court agreed that the Montana tax meets the first two parts of the test. The majority argued that it is virtually impossible for a resource severance tax to discriminate against interstate commerce even if it is largely paid by out-of-state consumers. Further, the majority emasculated the fourth part of the test by holding that the only requirement is that the measure of the tax be reasonably related to the extent of the contact between the business and the state. Therefore it affirmed the decision of the Montana courts. This effectively destroys constitutional protection. The minority, led by Justice Blackmun and including Justices Powell and Stevens, desired to hold the states to a much higher standard in fairly relating the tax to the services provided by the state.

One may reasonably interpret the dissenters as trying to maintain the fourth part of the *Complete Auto* test. According to the majority's views there is essentially no constitutional protection from extremely high severance taxes as long as the proper forms are observed in defining them. However, at many points the court mentioned the power of Congress to limit state severance taxation.

The high severance tax rates set by states with large energy production have produced a political reaction among the have-not states. A coalition of congressmen from energy-consuming states has formed. A member of this coalition, Rep. Robert W. Edgar, Democrat from Pennsylvania, has called the states imposing high severance taxes the "United American Emirates," drawing a parallel to OPEC's attempt to monopolize oil production.

This activity has led to the introduction into Congress of several bills designed to limit state severance taxes to 12.5 percent of value (Fish 1982, p. 1032). Note that this would be an imperfect solution because there are resources that are relatively costly to exploit, such as California heavy crude oil, where a 12.5 percent tax (on the gross value of the resource) would be very burdensome, while such a tax on Montana coal or Texas crude would not present such serious problems.

3.2 Theoretical Analysis in the Economic Literature

Charles McClure (1969) made an early contribution to the theory of tax exporting, looking at tax incidence in a simple general equilibrium model. This model assumes that labor is immobile, capital is mobile, each state is perfectly specialized in producing only one good, and residences of workers, consumers, and capitalists are fixed. The model also ignores tax exporting that arises from the deductibility of state and local taxes from the federal corporate income tax. Leaving aside the obvious exporting from taxing commuters, McClure's simple model indicates that general consumption taxes are not exported, while production taxes are exported if they raise the market price of the good produced by the taxing state or if the nontaxing state is a net creditor. The creditor status of the nontaxing state matters because a production tax reduces the rate of return on investment in the nontaxing state. McClure draws clear policy implications:

> from the standpoint of efficient fiscal federalism, (consumption taxation) may be a preferred form of taxation, except in cases where government expendi-

tures lower costs of production. This implies that continued state use of retail sales and income taxes may be far better than a shift to taxes tied more closely to production. . . . (p. 478)

Todd Sandler and Robert Shelton (1972) expanded the work of Mc-Clure by including spillovers of public goods across states and by explicitly considering the issue of stability of the nation (that is, economic incentives for any states to leave the union). They showed that both spillovers and tax exportation can lead to inefficiencies. Further, they stressed the problem of tax harmonization. That is, it is in the interest of all parties to limit tax exportation, but it is not in the interests of any one tax-exporting state unilaterally to limit its own taxation. Given this, they expect that it should be possible to obtain agreement from all states on a constitutional amendment to limit tax exportation.

McClure's model was generalized with respect to production taxes by Shelby Gerking and John Mutti (1981). They allowed one of the states to be incompletely specialized. This undermines the crisp results of Mc-Clure. McClure had shown that the nontaxing state must become worse off if production taxes are used, because its terms of trade are made worse. Allowing for incomplete specialization complicates this argument because the impact on the terms of trade depends on the relative factor intensities of all industries.

An optimal taxation approach has been taken by Roger Gordon (1973). His model may be considered as long run, since he allows for mobility on the part of labor and consumers as well as capital and public good spillovers. Further, he assumes that states set taxes and expenditures to maximize the expected utility of the state's current residents. He allows for incomplete specialization in production by the states. In comparing results when states make independent decisions, assuming that taxes and expenditures of other states will not change (in game theory terms, a Nash assumption), he finds many sources of externalities and thus inefficiency. He notes that spillovers may lead to particular taxes being set too low (if high taxes cause economic activity to migrate to other states) or too high (if the tax itself is exported). For similar reasons the total level of taxation may be either too high or too low. If the federal government believes that the total level of taxation would be too low, it can compensate by allowing the tax deductibility of state taxes (as it does). Other alternatives also observed are revenue sharing and block grants. On the other hand if specific taxes appear to be high, Gordon suggests that the federal government can either prohibit such taxes or legislate a maximum rate. He relieves that state severance fees and perhaps state corporate income taxes may be examples of this. In common with McClure, Gordon notes that taxing factors where they are owned and goods where they are consumed reduces the externality problem.

In a recent paper Dennis Olson (1984) generalized the preceding approaches by introducing a pure intermediate good that is taxed in a federal system. As one would expect, this weakens the positive results. Depending on factor intensities it is possible, though unlikely, that taxation of the intermediate good would raise the output of at most one final good and raise the rewards of one factor. The most probable result, of course, is that all final goods would be produced in smaller amounts and all factor prices would fall. Olson applies this model to the specific problem of severance tax exportation and finds that the energy-producing states may benefit by tax increases but the country as a whole would lose. Therefore "some type of federal restraint limits imposed on state severance taxes might be necessary to protect national welfare" (p. 122).

In much of the literature on taxation, there is the clear implication that it is beneficial to tax immobile factors, such as natural resources, because this would cause little or no geographic distortion of economic activity. After all, the mine or the oil well can hardly migrate across state or local boundaries. In a recent essay, McClure (1984) has challenged this conventional wisdom. He argues that geographic variation in the presence of resources allows some states to adopt tax and expenditure policies that lead to inefficient migration of mobile factors. The most striking example is the recent attempt of Alaska to pay a rebate to its residents based on its oil severance tax revenue. Needless to say this would lead to inefficient migration of labor and capital to Alaska. McClure shows that efficient taxation of such resources can be achieved only by federal taxation or extensive revenue sharing among states. He apparently expects Congress to limit state severance taxation, though he argues that the states have other instruments to achieve the same purpose, so that federal limits to state taxing power should be broader than severance taxes (pp. 157–58).

3.3 Empirical Findings

McClure (1967) produced early empirical estimates of tax exporting in 1962, based on the same theoretical framework as his theoretical analysis. His analysis has long-run and short-run components, and these components adopt different assumptions regarding capital mobility and the ability of states with market power in particular industries to raise national prices. The issue of shifting the tax burden forward to consumers is handled by assuming that industries dominated by a particular state can shift 60 percent of the burden to national consumers in the short run but only 40 percent in the long run. A state is assumed to dominate its industry if it produces 40 percent of the value added for a nationally marketed good. This seems likely to overstate forward shifting. The analysis takes careful account of the residence of capitalists and workers and interactions through the national tax system.

The estimated rate of tax exportation varies somewhat across states and between short and long run and greatly across the type of tax applied. As expected, general consumption taxes are not exported very much, ranging in the short run from 14.9 percent in Indiana to 33.4 percent in Nevada (since so much of the tax is related to gambling). At 27.2-percent exporting, California is slightly higher than average. In the long run export rates are slightly lower than in the short run. Selective sales taxes are exported somewhat less than general consumption taxes, but naturally there is more variation in their export rates.

Business taxes are exported to a much greater extent than general consumption taxes. For corporate income and franchise taxes, in the short run nonresidents pay a high of 91.0 percent for Delaware. The low, North Carolina, is still 60.7 percent. Slightly lower than average, California exports 70.0 percent. In the long run less forward shifting is possible, and capital becomes mobile, so less of the tax burden is exported. But exportation still ranges from 81.9 percent in Delaware to 44.4 percent in South Carolina. California exports 46.1 percent, again slightly below the average. Property taxes are exported to a much lower extent, ranging from 29.9 percent for Arizona to 11.0 percent for the District of Columbia in the short run. Arizona is high because taxes on railroads, tourism, and mining are assumed to be highly exported. California is a bit below the mean, with 19.0 percent exported. In the long run exportation is slightly lower. In conclusion McClure stresses that "for the bulk of the taxes on mineral property or severance, railroad property, and manufacturing property, there is little chance that exported benefits equal exported taxes. This being the case, these taxes are clearly sources of inefficiency" (p. 73).

With some changes and limitations this study has been updated by William Morgan and John Mutti (1985) to the year 1980. The general pattern of results shown in table 3.1 is the same, with business taxes exported far more than consumption taxes. However, the overall level of exportation is found to be much higher, even though the authors, by assumption, ignore the possibility of forward shifting through raising

TABLE 3.1
Estimated Tax Exportation of State Corporate Income and Property Taxes, 1980
(dollars in millions)

	Taxes	% Exported		Taxes	% Exported
Alabama	$169	98.8	Missouri	378	97.9
Alaska	613	100.0	Montana	135	100.0
Arizona	271	98.5	Nebraska	177	99.4
Arkansas	145	100.0	Nevada	54	100.0
California	3,841	88.3	New Hampshire	162	99.4
Colorado	341	98.8	New Jersey	1,352	95.9
Connecticut	543	97.2	New Mexico	79	100.0
Delaware	65	100.0	New York	4,264	87.8
District of Columbia	138	100.0	North Carolina	494	98.2
Florida	743	92.6	North Dakota	75	100.0
Georgia	475	98.3	Ohio	1,050	96.0
Hawaii	86	100.0	Oklahoma	180	98.9
Idaho	81	100.0	Oregon	449	98.9
Illinois	2,152	94.3	Pennsylvania	1,533	94.5
Indiana	558	98.2	Rhode Island	131	100.0
Iowa	393	99.2	South Carolina	237	99.6
Kansas	332	99.4	South Dakota	50	100.0
Kentucky	220	98.6	Tennessee	497	98.6
Louisiana	360	98.9	Texas	1,156	94.9
Maine	118	100.0	Utah	103	100.0
Maryland	347	98.0	Vermont	56	100.0
Massachusetts	1,133	96.5	Virginia	402	97.8
Michigan	1,184	96.6	Washington	134	98.5
Minnesota	667	98.4	West Virginia	90	100.0
Mississippi	128	100.0	Wisconsin	613	98.5
			Wyoming	59	100.0
U.S. total				29,013	94.5

SOURCE: Morgan and Mutti (1985, pp. 196–98).

NOTE: Total includes part of value added tax estimated to fall on capital.

prices to out-of-state consumers and also do not directly consider mineral severance taxes where this is especially likely to occur.

Morgan and Mutti find that the average rate of exportation for taxes on corporate and noncorporate business (excluding severance taxes) is 83 percent, while it is only 24 percent for household taxes. The average rate for taxes on corporations is 94.5 percent. California exports 88.3 percent of its taxes on corporations. In many small states with few stockholders (for example, Alaska, Delaware, West Virginia), approximately 100 percent of the taxes on corporations is exported. The main reason for the discrepancy in exportation rates between the McClure and Morgan and Mutti studies is the enormous increase in federal taxation from 1962 to 1980. The average marginal federal income tax rate was estimated to be about 18 percent in 1962 (McClure 1967, p. 54). By 1980 it had doubled to about 36 percent (Morgan and Mutti 1985, pp. 200–201). This difference alone accounts for 18 percent more exportation in 1980, enough to make up the difference between the two estimates.

The authors also compute taxes imported by each state. The mirror image of exportation, importation occurs through the federal tax system and ownership of stock in corporations that are taxed in other states.

By subtracting tax imports from exports, the authors calculate the net tax exportation position of states. Ignoring mineral severance taxes, the overall rate of net tax exportation is found to vary between 62.8 percent for Alaska and − 30.0 percent for Florida. Most of the variation is due to differences in the proportion of stockholders resident in the state relative to the amount of taxes raised by taxing corporations. Florida is a big loser because its elderly residents, of course, own a disproportionate amount of corporate stock. Morgan and Mutti discuss the complications of long term contracts and rail or pipeline market power that make short-run analysis of state energy severance taxes difficult. But they note that these taxes are important for energy-producing states and if included, severance taxes would drastically change the net exportation position of these states.

From these very high rates of tax exportation of taxes on corporations, one might think that states would rely entirely on them to raise revenue. It looks like a free lunch. In the short run relying on exported business taxes would work. What restrains states is the long-run mobility of capital. High business taxes lead to a flight of capital from the state. This reduces the demand for labor and land. Wages and rents then decline enough to stop the exit of capital. But in the process the incidence of taxes on business has shifted to the immobile factors of labor and land in the taxing state. For most businesses that operate in national markets, this means that in the long run the only source of tax exportation is through the federal tax system. If so, the long-run rate of exportation would be only about 36 percent (Morgan and Mutti 1985, pp. 200–201).

States that are able to tax minerals in some way can largely avoid this problem, since a mineral deposit can hardly move to another state. And it is likely to be owned, either outright or in effect, by corporations whose shareholders live elsewhere. Therefore a severance tax on mineral production would be paid largely by owners of the resource. To the extent that output would decline, some part of the burden would be shifted to labor. Most importantly, the short-run and long-run degree of exportation would be approximately identical. Morgan and Mutti believe that on average, about 67 percent of state severance taxes is exported (1985, p. 206). This is about twice as high as the long-run rate of exportation for business taxes on mobile resources.

Further, for minerals like western coal, some states have monopoly power so that part of the tax can be exported in the form of higher prices for the mineral. Thus these states export even more of the burden of taxes. This explains the extremely high 30-percent severance taxes on coal production in Montana. This type of forward shifting of the burden is impossible for oil production, since the United States is a major net importer of crude oil and no state's output of crude oil is a very large proportion of the world crude oil market. Thus it is not surprising that the highest severance tax rate for oil is 12.5 percent in Alaska, far less than 30 percent.

Whether or not taxes can be shifted forward to consumers, state severance taxes are the most tempting for states to force out-of-state residents to pay their taxes, even in the long run. This is one reason they have received so much attention in law, politics, and economics.

Chapter 4

Taxation and Petroleum Supply

4.1 Introduction

Changes in the revenues producers expect to receive from extracting crude oil, natural gas, and other natural resources will alter their incentives to develop and produce these resources. In general, either a reduction in market price or an increase in the tax liability associated with production will lower profitability and therefore the extent of resource development. A priori reasoning will not, however, carry one very far beyond this simple generalization. Quantitative statements regarding the response of resource development to changes in revenues require empirical analysis specific to the resource in question.

In the present chapter we survey a range of models developed to measure the price responsiveness of crude oil and natural gas supplies. The intent of this exercise is to obtain guidance in modeling crude oil supply in California. In addition we examine the effects of changes in the market price of crude oil and changes in crude oil severance tax rates on net revenues from production. This analysis is necessary to permit the supply effect of a severance tax to be inferred from a model of the response of crude oil supply to market price.

4.2 A Survey of Oil and Gas Supply Models

Since 1970 a variety of crude oil and natural gas supply models has been developed. Given the purpose of this study, those aspects of supply models that determine sensitivity of supply to the after-tax marginal revenues of petroleum developers are of particular interest. To direct attention to the distinguishing features of a broad array of supply models, a two-part categorization consisting of geologic-engineering models and econometric models is useful.

4.2.1 Geologic-Engineering Models

This class of models is characterized by the maximal use of geologic and engineering information to forecast total reserve levels, exploration rates, production schedules, and so forth, under various price-cost conditions. Exemplary of this class of models are Attanasi et al. (1980), U.S. Geologic Survey (1981), Arps and Roberts (1958), and Kim and Thompson (1975). Rather than survey each model in detail, the following discussion concentrates on the model developed in Attanasi et al. (1981) to describe the general geologic-engineering approach. The discussion of other models in this category is then confined to a brief account of their distinguishing characteristics.

The model formulated by Attanasi et al. (1981), and more fully described in U.S. Geologic Survey (1981), was intended to permit predictions of undiscovered potential reserves, and associated exploration, development, and production costs, as a function of wellhead price and the rate of return required by developers. The authors' general methodology was applied to the Permian basin of west Texas and southeastern New Mexico, a mature and highly developed petroleum-producing region.

The general approach taken by these authors may be broken down into three components. The first is a discovery-process model that uses historical information on exploratory drilling and field discoveries to estimate the number of undiscovered oil and gas fields in this region by size and depth category. Additional geologic and engineering data were used to predict such field characteristics as the ratio of primary oil fields to total oil fields, ultimate recovery of oil versus gas, and production per well. The authors also formed estimates of additional production available from use of secondary recovery for each size and depth category.

The second component of the Permian basin model is a discounted cash flow analysis for each field category. This is accomplished by specifying decline rates and annual operating costs for each field type. Discounted cash flows for each category of field were then computed for a variety of price levels and discount rates.

To determine total exploration activity, and the proportion of all undiscovered fields that will eventually be discovered and developed, the authors compare the expected present value of net revenue from production to exploration costs. Thus using historical data on finding rates and exploratory drilling costs for each size-depth range, Attanasi et al. (1981) compute the expected cost of finding a representative field in each category. If this finding cost is less than the expected present value of production from discovered reserves, then the field is economic to find, and according to their central assumption, it will eventually be found and developed. Since discounted cash flows depend on an assumed price and rate of return, the result of their analysis is a schedule of fields, and associated reserves, production, and exploration-development activity that will eventually be developed at each price–rate of return combination included in their analysis. In a check of their model Attanasi et al. (1981) use it to estimate reserves in fields already discovered. Overall their model appears to provide reserve estimates that are very close to estimates based on engineering studies of individual fields. This check does not, however, verify their estimates of discovery costs for various size-depth classes.

The geologic-engineering approach, as represented by Attanasi et al. (1981), has much to recommend it. Most importantly perhaps it uses geologic information in a way that rules out making predictions regarding future supplies that are grossly unrealistic in light of the record of geologic evidence. Moreover, it is readily amenable to the addition of any supplemental geologic information that may be available. This is not in general true of econometric models of oil and gas supply. These models largely extrapolate past relationships between reserve additions, production, and so forth, and such economic factors as prices and interest rates. Unless suitably adjusted, econometric projections may not provide an accurate guide to the future if geologic attributes of undiscovered fields differ significantly from those represented in the sample. Moreover, this lack of similarity is likely, since the most economically attractive fields tend to be discovered first.

Another attribute of geologic-engineering models, which might or might not be counted as an advantage, is that they guarantee the sensitivity of reserves and production to such economic factors as prices and interest rates. Clearly, if prices are higher or interest rates lower, the set of potential fields that are economic to develop is expanded.

A major disadvantage of the approach taken by Attanasi et al. (1981) is that it does not automatically provide time paths for exploration activity, reserve additions, and production. Rather it provides only cumulative (infinite horizon) totals for these variables. This is particularly unfortunate for policy analysis. To translate these cumulative estimates into rates per period, it would be necessary to introduce lags, rigidities, the formation of expectations, or other factors into the analysis. Without such constraints these models may project implausibly large changes in exploration, production, and so forth, over short time periods (see Epple 1985, p. 144). Unless the resulting time paths incorporate evidence on the past responsiveness of drilling, reserve additions, and production to changes in the economic environment, there is no guarantee that the time patterns constructed would be plausible from an economic perspective.

Other supply models included in the geologic-engineering category are Arps and Roberts (1958), Kim and Thompson (1975), White (1981), and Davis and Harbaugh (1981). Of course none of these models adopts precisely the methodology used by Attanasi et al. (1981). All share the similarity, however, that supply forecasts are based on the simulation of optimizing decisions in the context of relatively extensive information on geologic characteristics of the resource. Arps and Roberts (1958) developed the discovery-exploration relationship used by Attanasi et al. (1981) but did not pursue the sensitivity of reserves to economic factors in any detail. Kim and Thompson (1975) develop a geologic model of the sensitivity of reserve additions to exploratory drilling and use a decline rate approach to specify production. A time path for exploration is then derived from a model of investment in exploratory drilling capacity. Given the decline rate characterization of production, this suffices to determine the time path of supply for a given price regime.

The models developed by White (1981) and Davis and Harbaugh (1981) are similar to one another in that both promote the use of simulation and Monte Carlo techniques to model how exploration provides information on reserves. These differ from other studies surveyed in this section, however, in that they focus exclusively on the exploration process.

Finally, Camm et al. (1982) present a simple model of production scheduling from existing (discovered) fields. In their framework only the initial production capacity (which determines a production decline rate) and shut-in date are choice variables for the operator. Once the investment decision is made, only the shut-in date can be varied in response to changes in costs or net of tax prices. Their production-scheduling model is then used to examine supply responses to a severance tax. These short-run effects are quite naturally shown to depend on the decline rate for the field in question, the pretax shut-in date, and the age distribution of wells in the field. No significant analysis of exploration or development effects of such taxes is offered.

4.2.2 Econometric Models

The set of supply models classified as econometric all share the common feature that the responsiveness of production, drilling, and reserves to economic variables is estimated from time series data. Given the economic similarity, and frequent joint production, of oil and gas resources, both oil and gas supply models are included in the following discussion. Among the numerous oil and gas supply models now represented in the literature, only those widely cited and discussed are outlined. These are Erickson and Spann (1971); Erickson et al. (1974); MacAvoy and Pindyck (1973); Epple (1975); and Epple (1985). Several other supply models or supply estimates are briefly mentioned at the end of the discussion.

Perhaps the best representative of this class of supply models is MacAvoy and Pindyck (1973). Although their model was formally specified and estimated for natural gas, there appears to be no reason why it could not be readily adapted to examine crude oil. Their overall model consists of a system of simultaneous equations for different aspects of supply. Each equation is estimated from pooled time series and cross-section data, where regions in the cross sections are identified with dummy variables. The initial equation explains exploratory drilling as a function of expected revenue per well (price times average historical discovery size), average drilling cost in the preceding period, variance in average discovery sizes, and a set of regional dummy variables. The average size of discoveries is then specified to depend on the wellhead price, average drilling cost, the cumulative number of wells drilled, and the same set of regional dummies. With discovered reserves identifiable from the first two equations, reserve extensions and revisions are modeled to follow lagged discoveries, lagged reserve revisions, and total exploratory drilling. Production from reserves is in effect modeled to depend on wellhead price, reserve levels, and regional dummy variables. A separate component of the model consists of a set of wholesale demand equations.

The models developed by Erickson and Spann (1971) and Erickson et al. (1974) are similar, since both focus attention on the determinants of reserves for natural gas in the former and crude oil in the latter. Erickson and Spann (1971) are similar to MacAvoy and Pindyck (1973) in that separate equations are estimated for individual components of the reserve addition process. In particular their model separates equations for wildcat drilling, the success ratio on exploratory wells, and the average discovery size. In general these variables are taken to depend on prices, geologic variables, and lagged values of dependent variables. The Erickson et al. (1974) model of oil reserves specifies that a firm's desired reserves depend on expected price, production restrictions (prorationing), the opportunity cost of capital, and tax parameters. Novel features of this model are the explicit treatment of price expectations and the notion that reserve investment decisions represent lagged responses to differences between desired and actual reserves. Unlike MacAvoy and

Pindyck (1973), however, neither of these reserve models is extended to estimate actual supply responses. It should be noted that Pindyck (1974) and Neri (1977) have performed comparative evaluations of the preceding models.

Epple (1975) developed a supply model that focuses attention on the supply of oil-bearing properties (mineral rights) as the primary determinant of long-run supply. The exploration process is treated as an ordinary production activity in which outputs of oil and gas discoveries are produced with inputs of labor and capital (drilling) and mineral rights. Static profit maximization by explorers yields a derived demand for mineral rights, and the supply of mineral rights is taken to depend on the rent per unit received by mineral rights owners. The direct application of this approach to obtain a long-run supply function for discoveries requires data that typically are not reported, for example, unit prices for mineral rights and new petroleum discoveries. Discovery prices are generated by a zero (normal) profit assumption for the exploration activity, and the supply function for mineral rights is inferred from other parameters in the model.

The supply model developed by Epple (1985) reflects an attempt to derive an econometric model of supply from an explicit formulation of the producer's optimizing problem. In particular, Epple (1985) concentrates the modeling effort on dynamic decision making in the context of uncertainty regarding future prices and costs and interrelationships between current production levels and future production costs. This is an ambitious goal, and it forces certain compromises on the structure of the model. The econometric analysis is confined to estimating a model of reserve additions. Field development and production are handled by relatively simple mechanical rules (for example, production each year is a constant fraction of reserves). The author also incorporates a detailed analysis of tax provisions in the producer's optimizing model.

Several other oil or gas supply studies appear in the literature but are not discussed because the analysis they present is either similar to or more limited than that found in papers already cited. Examples are Mancke (1970), Erickson (1970), Khazzoom (1971), Cox and Wright (1976), and Uhler (1976).

When the purpose of energy supply analysis is to predict the effects of price, tax, or cost changes, the econometric analysis of time series data has much intuitive appeal. In particular the lags, rigidities, and uncertainties that prevent immediate and complete responses in investment and production decisions are implicitly represented in such models, at least to the extent that the historical pattern of rigidities is indicative of the future. As noted earlier, simple application of present value-maximizing geologic-engineering models, without the addition of such constraints, can yield supply and investment effects that are implausibly dramatic.

Econometric models of oil and gas supply have so far made little use of geologic information in specifying functional forms or imposing restrictions on parameters, however. An advantage of the geologic-engineering approach is that it readily admits the use of geologic information on the size and depth distributions of undiscovered fields and the likely costs of finding and developing them. The incorporation of such information would appear a promising direction in which future econometric models might be extended. This refinement would require structural equations in these models to be specified to allow the use of such information. In practice such extensions may be hindered by aggregation problems, for example across fields in different size and depth categories, in which case a multiequation approach and a very extensive (and possibly unavailable) data base may be required. A potentially attractive alternative would be to estimate certain facets of the overall supply response, for example cumulative exploration or reserve additions, from a geologic-engineering model and then use an econometric approach to estimate time paths.

Finally, with the exception of Epple (1985), econometric supply models have paid little attention to the optimization problem that presumably underlies them. That is, the theory and estimation of exhaustible resource supply are seldom connected in any significant way. Similarly, the difficulty of forecasting price and cost surely plays a significant role in actual investment and production decisions, yet it is generally overlooked when econometric specifications are adopted. Epple's (1985) efforts in these areas are pathbreaking, but they also demonstrate that the cost of such rigor is likely to be a substantial increase in the complexity encountered when formulating and estimating econometric models of energy supply.

4.2.3 Conclusions

Estimates reported in several studies make it possible to compute implied supply elasticities for oil and gas. Given the intertemporal nature of petroleum production, supply responses to price changes can be stated only with respect to a given time period for production. Most supply elasticity estimates pertain to reserves and thus relate to an infinite production horizon. A sample of elasticity estimates is presented in table

TABLE 4.1
A Summary of Crude Oil Supply Elasticity Estimates

Author and/or Institution	Elasticity Estimate	Description
U.S. Geologic Survey (1980)	1.14 1.0	Elasticity of reserves price range $22–$29 price range $29–$37
Mancke (1970)	1.0–2.0	Long run; presumably refers to reserves
Erickson and Spann (1971)	0.9	Elasticity of reserves with respect to natural gas price
U.S. Cabinet Task Force (1970)	1.0	From survey of oil producers
U.S. Energy Information Administration (1978)	0.73–1.25[a]	Price range of $21–$28; specific to California onshore production

[a] Five-year production horizon.

4.1. Surprisingly, perhaps, these estimates are all of roughly the same magnitude despite the fact that they generally pertain to supplies from different regions and represent a wide range of underlying methodologies.

Received supply models generally enable forecasts of only long-run production responses to price changes. Most commonly estimates are provided for total reserves under alternative price regimes and give little attention to the timing of production from these reserves. The time profile of these economic impacts is, however, often of paramount importance to policy makers. This consideration was of central concern in the development of the supply model reported in chapter 5.

The general framework we adopted for analyzing the supply effects of a severance tax on crude oil in California uses a mixture of econometric and geologic-engineering methodologies. The primary econometric component is our representation of the effects of the tax on reserve additions. Although the geologic-engineering method has several strong points in this area, it fails to provide an endogenous predicted time path for exploration and development activity, and this was a major factor in our choice of an econometric approach to this aspect of petroleum supply. Our reserve additions component treats separately the completion of new producing wells and reserve additions per well. In this sense it employs a methodology similar to that developed by Erickson and Spann

(1971) and MacAvoy and Pindyck (1973). Time series data are used to identify the relationship between field prices and exploratory and development drilling. A lagged adjustment model of price expectations provides a framework in which the drilling responses to price changes are phased in as price expectations are revised. The method used to translate new wells into reserve additions differs somewhat from that used in other models because it relies on forecasts of initial production per well and production decline rates for new producing wells.

To estimate the effect of a severance tax on production from existing wells and future reserves, we rely on a framework that is in the spirit of the geologic-engineering approach. It was earlier noted that received econometric models have paid little attention to the price responsiveness of time paths of production. Further, most geologic-engineering supply studies have sought only to examine changes in reserves. Camm et al. (1982), however, provide a methodology that combines a geologically fixed production decline schedule with simple cash flow analysis to determine the price responsiveness of production from reserves, on a field by field basis. Since this framework provides explicit production time paths as functions of prices, and uses data readily available for California, it provides an ideal starting point for our analysis. Thus the production component of the supply model presented in chapter 5 is an adaptation of the approach developed by Camm et al. (1982).

4.3 Tax Interactions and Supply Responses

When studying the supply effects of a change in petroleum tax policy, one typically cannot draw direct inferences from historical responses to tax rate changes, for example by inspecting a tax rate coefficient in a regression equation. Tax rates are changed only infrequently, and in the case of interest, here no significant severance tax has ever been imposed on California crude oil. Consequently one must infer supply responses to tax changes from estimated responses to price changes. That is, one must be able to translate a given change in the tax rate into an equivalent price change and then infer the supply effects of taxation from the estimated responsiveness of production to price.

Translating tax changes into equivalent price changes is complicated by the presence of a multitude of taxes on petroleum operations and by the fact that these taxes interact with one another through the definitions of tax bases. For example, state severance taxes are deductible from taxable corporate income at the federal level. Thus a $100 million payment of state severance taxes will lower the firm's federal tax liability by

$46 million if the firm's effective marginal corporate tax rate is 46 percent. Similarly, state severance taxes partially offset federal windfall profits tax payments. Severance taxes paid on the value of oil that exceeds its value in base period prices (that is, the so-called windfall) are deductible from the base of the windfall profits tax. Even property taxes can interact with severance tax payments if assessed property values are based on the economic value of reserves. If a severance tax renders some reserves uneconomic to recover, then the firm's property tax liability will be reduced accordingly.

For all of these reasons, identifying a price change that is equivalent to a given severance tax rate is rather complicated. Our detailed analysis of this issue is presented in appendix *B*. This analysis leads to a very simple conclusion, however. To a first approximation, the imposition of severance tax at *x* percent will have the same effect on the firm's energy supply decision as an *x*-percent reduction in the wellhead price of crude oil. In the absence of the windfall profits tax (WPT), this simple equivalence holds precisely; with a windfall profits tax, however, the equivalence is only approximate. The result can be grasped intuitively from the stylized example presented in table 4.2. Consider, first, cases *A* and *B*

TABLE 4.2

Tax Liabilities and the Shut-in Decision: An Illustrative Example

	Baseline	6% Severance Tax	6% Price Reduction
1. Revenue/bbl.	$25.00	$25.00	$23.50
2. Severance tax	0.00	1.50	0.00
3. WPT payment	1.50	1.41	1.35
Net income in absence of corporate income tax if operating cost is			
4. $22.00/bbl.	$1.50	$0.09	$0.15
5. $23.00/bbl.	0.50	(0.91)	(0.85)
6. $24.00/bbl.	(0.50)[a]	(1.91)	(1.85)
Net income after 50% corporate income tax payment if operating cost is			
7. $22.00/bbl.	$0.75	$0.05	$0.08
8. $23.00/bbl.	0.25	(0.45)	(0.43)
9. $24.00/bbl.	(0.25)	(0.95)	(0.93)

NOTE: Assumptions: Severance tax rate is 6%; WPT base price is $20.00/bbl., and tax rate is 30%; severance and WPT payments are deductible from income tax.

[a] Figures in parentheses denote losses.

reported in the first two columns of figures in table 4.2. Here revenues and tax liabilities for a hypothetical property are shown under a uniform set of assumptions regarding all relevant factors except the presence of a severance tax. Price per barrel is set at $25.00, and the details of the assumed WPT and corporate income levies are explained in the accompanying note. In this example entries 4–6 show the operator's net revenue in the absence of any corporate income tax liability under various operating cost assumptions. As oil wells age, production rates decline and operating costs per barrel rise. As case A shows, the hypothetical well could operate profitably at costs exceeding $23.00/bbl. but would be shut in before costs rose to $24.00/bbl. Case B represents the finances of the same property after the introduction of a 6-percent severance tax. As entries 4–6 demonstrate, the property would now be shut in before operating costs rose to $23.00/bbl. This is an example of premature shut-in due to the imposition of a severance tax. Note also that the severance tax has reduced the firm's WPT liability somewhat. As a consequence of this offset, the firm's net tax liability has risen by only $1.41 rather than $1.50 as a result of the severance tax. (The $1.41 is computed as the difference in net revenue before income tax under cases A and B.)

Entries 7–9 report the firm's net income in the case where a 50-percent corporate income tax is assessed. (This tax may be considered a combined state-federal rate.) These figures are quite simply half as large (in absolute value) as those in items 4–6. It follows directly, then, that the shut-in effect of the severance tax is the same with or without corporate income tax. In the presence of an income tax the well is again shut in after costs have risen above $23.00/bbl. but before they reach $24.00/bbl. in case A. Entries 7–9 for case B show that the severance tax induces shut-in before costs rise to $23.00/bbl., as was the case without an income tax. It should therefore be clear that the shut-in decision is independent of whether or not an income tax is levied. Nevertheless it can be seen that the WPT and the income tax have cut the net tax liability associated with the 6-percent severance tax. Comparing cases A and B in any of the net income entries 7–9 reveals that the firm's net severance tax liability is only $.70/bbl. rather than the $1.50 obtained by applying the statutory rate to the gross revenue. However, this deductibility in no way mitigates the impact of the 6-percent severance tax on the shut-in decision.

Case C is reported in the third column of figures and was included in table 4.2 to demonstrate that a 6-percent severance tax has approximately the same effect on the shut-in decision as a 6-percent price reduction regardless of the presence of WPT and income taxes. Comparing cases *B* and *C* reveals that net incomes in the two situations (*B* with a price of $25.00/bbl. and a 6-percent severance tax; *C* with a price 6 percent lower but no severance tax) are almost identical. Consequently the shut-in decision would be almost identical in the two cases, that is, a 6-percent severance tax affects production from existing wells to almost the same degree as a 6-percent price reduction. As the example shows, the severance tax actually causes a somewhat larger net income reduction than the price cut and hence would also have a somewhat larger impact on production. This occurs because only part of the firm's severance tax payment is deductible from its WPT liability (that portion that is levied on the so-called windfall); a price reduction on the other hand is fully deductible from the WPT.

The discrepancy reflects a difference in the effects of severance taxes and price reductions on the base of the WPT. Deductibility provisions in the WPT guarantee that a 6-percent severance tax will reduce the firm's taxable windfall by exactly 6 percent. As shown in appendix *B*, however, a 6-percent price reduction will cause the firm's WPT liability to fall by more than 6 percent; hence its after tax marginal revenue will fall by less than 6 percent. One can correct for this discrepancy by basing the supply effects of a severance tax on the price responsiveness of supply estimated for the period prior to introduction of the WPT. When this is done the simple equivalence between a 6-percent price reduction and a 6-percent severance tax is maintained. This point is demonstrated in more detail in appendix *B*, and the supply model presented in chapter 5 is formulated accordingly so that this equivalence is preserved.

There are two circumstances where a given severance tax rate can interact with other aspects of the firm's operating costs to magnify substantially the impact of the tax on production. The first of these concerns liability for severance tax payments on royalty production. Suppose the operator of a field is wholly liable for severance tax payments on all oil produced and the operator is paying royalties to those who own the property. If the severance tax rate is 6 percent and the royalty owner's interest is one-sixth, then the effective tax rate on the operators' net revenue is raised to 7.2 percent [6 percent $\times$ (7/6)]. A more detailed demonstration of this point appears in appendix *B*.

Typical leasehold agreements specify that the royalty owner is liable for severance taxes on royalty production, and in such cases the compounding of effective severance tax rates just explained does not occur. There are, however, properties in Kern County which specify that the operator is liable for severance taxes on all production. The extent of such arrangements and the royalty rates applicable to them are unknown.

Thus the supply model presented in chapter 5 ignores this consideration and implicitly assumes that operators are liable only for severance taxes levied on their economic interest in the field.

The second way in which the production impact of a given severance tax rate could be compounded is if the lease fuel used in thermal recovery projects is subject to the tax. The potential magnitude of this effect is readily apparent from an example. Suppose the operator must burn one barrel of fuel for each three that are produced for sale. If a 6-percent tax is paid on the wellhead value of all four barrels, then the result is an effective 8-percent tax on the three that are actually sold. To obtain the effective tax rate on *net* output in these cases, the nominal rate must be scaled up by the ratio of total to net production. An efficient steam injection program can burn as little as one barrel out of every four barrels produced, so that the total to net production ratio is 1.33. This deteriorates to about one barrel burned from each two produced at the economic limit, for a ratio of 2.0. For these projects a 6-percent tax applied to gross production would amount to 8 percent of net production initially and would eventually rise to 12 percent. The importance of this is evident from the fact that enhanced oil recovery in California, which is almost exclusively by steam injection, was credited with about 28 percent of total California crude oil production in 1980 (see chapter 2).

Recent severance tax proposals in California have specifically excluded lease fuel from taxation. For this reason the magnification of the severance tax rate that taxation of lease fuel would cause is not incorporated in the supply analysis that follows.

Appendix B

Tax Interactions and Supply Responses: Analytical Details

This appendix presents technical details of the argument used to translate a given severance tax into an equivalent change in the wellhead price, as required to estimate the drilling and production effects of a severance tax from the crude oil supply model developed in chapter 5. The impact of a severance tax on decisions to explore and drill will depend on the way the tax affects net revenue from production in a given field. The effect of a severance tax on the shut-in date for a given field will depend on the way the tax affects the marginal revenue from producing an additional barrel of oil. The supply model presented in chapter 5 specifies that production from existing fields continues so long as net cash flow is positive; that is, so long as marginal revenue net of taxes exceeds the average cost of production.

Our notation and analytical approach are similar to Camm et al. (1982, pp. 145–59), but our conclusions differ in certain respects. We find that to a first approximation, imposition of an x-percent severance tax will have the same effect on the firm's shut-in decision as an x-percent reduction in the wellhead price. Moreover, in the absence of the WPT, this simple equivalence would also apply to the exploration and development decisions. In the presence of a WPT, however, an x-percent severance tax will have an effect equivalent to a somewhat larger percentage price reduction. The preceding conclusions apply to severance taxes as commonly applied. There are, however, circumstances that can substantially magnify the impact of a severance tax on exploration, development, and production decisions. These considerations are explained briefly at the end of this appendix.

The conclusions just outlined result from the interaction among taxes that arises because some taxes are deductible from the bases of other

63

taxes. To analyze these interactions it is convenient to adopt the following notation:

p Wellhead price of crude oil
q Quantity of crude oil produced per period
c Cost of production per period
I_f Federal corporate income tax per period
t_f Federal corporate tax rate
I_c State corporate income tax per period
t_c State corporate tax rate
R Royalty paid per period
r Royalty rate
S Severance tax per period
t_s Severance tax rate
W Windfall profit tax per period
t_w Windfall profit tax rate
p_b Base price for windfall profit tax
P Property tax per period
t_p Property tax rate
V Assessed value of reserves.

From the preceding definitions the firm's net revenue from production in a given field and a given period can be written

$$N = pq - c - I_f - I_c - R - S - W - P. \qquad (B.1)$$

Since I_c, R, S, W, and P are deductible from federal corporate income taxes, I_f can be expressed as

$$I_f = t_f(pq - c - I_c - R - S - W - P). \qquad (B.2)$$

State corporate taxes do not allow I_f to be deducted from taxable income, but typically do allow deductions for R, S, W, and P; thus

$$I_c = t_c(pq - c - R - S - W - P). \qquad (B.3)$$

Royalty payments are assumed to be a fixed proportion of a field's gross revenue in each period, so that

$$R = rpq. \qquad (B.4)$$

The operator is assumed to be liable for severance taxes only on the operator's economic interest in the field. This appears to be the most common arrangement in lease agreements, although in isolated instances the operator may be liable for severance taxes on the field's total output. We discuss these exceptional cases at the conclusion of this appendix but for the present specify that

$$S = t_s(1-r)pq. \tag{B.5}$$

A similar ambiguity arises in the operator's and royalty owner's liabilities for WPT. Here again the most common arrangement is for royalty owners to pay WPT on royalty production. In this case the operator's WPT liability is

$$W = (1-r)(1-t_s)t_w(p-p_b)q \tag{B.6}$$

which incorporates the provision that severance taxes paid on the windfall are deductible from the base of the tax. It should also be noted that the WPT contains a net income provision, a feature that is discussed later.

Finally, the operator's property tax liability is given by

$$P = t_p(1-r)V \tag{B.7}$$

which again specifies that the operator's tax liability is confined to his economic interest in the field. In general the assessed value of reserves in a given field will depend on price, costs, tax rates, and the volume of reserves in the field. In 1978, however, the voters of California adopted Proposition 13, and the result was to roll back values of taxable property in the state, including oil and gas reserves, to 1975 values. Under section 468 of title 18 of the California Administrative Code, assessed reserves may be revalued annually, but only to reflect "changes in the expectation of future production" (California Franchise Tax Board 1982, p. 49). As a consequence the assessed value per barrel of reserves is largely independent of price, tax, or cost conditions. The volume of reserves is, however, still influenced by these factors. Given the complications implied by this assessment rule, and the vagaries of actual assessment practices, it appeared futile to attempt to account for the relationship between assessed values, V, and all of the other tax, price, and cost parameters in the model. For this reason the following analysis treats V as fixed, and thus views the operator's property tax liability as a fixed component of the cost of production.

By combining equations (B.1)–(B.7) the operator's net income from a given field in a given period can be expressed as

$$N = (1-t_f)(1-t_c)(p(1-t_s)(1-r)(1-t_w(1-p_b/p))q - c). \quad \text{(B.8)}$$

Recall that our purpose is to identify price changes that are equivalent in their effects on the operator's exploration, development, and production decisions with a given change in t_s.

To sort out the rather complex relationship implied by equation (B.8), it is useful to first consider a simplified situation where only severance and corporate income taxes are present. Setting $t_w = 0$, the firm's operating revenue in this case can be written

$$N^0 = (1-t_f)(1-t_c)(p(1-t_s)(1-r)q - c). \quad \text{(B.9)}$$

It is clear from this expression that an x-percent decline in $1-t_s$ would have the same effect on the firm's profit opportunities from a given field in a given period as an x-percent reduction in p. In this simplified case, then, changes in p and t_s that satisfy the relationship

$$dp/p = d(1-t_s)/(1-t_s) \cong -dt_s \quad \text{(B.10)}$$

will leave N^0 unaffected where the approximation in equation (B.10) is valid for values of t_s near zero. Simply stated this result indicates that a t_s-percent severance tax will have exactly the same effect on the firm's profit opportunities, and hence its exploration, development, and production decisions, as a t_s-percent reduction in price. This equivalence is entirely independent of the presence or tax rates of state or federal income taxes.

Consider now the effect of introducing a WPT. We continue to ignore property taxes by treating them as a fixed component of operating costs. The firm's net revenue from a given field in a given period is yielded by equation (B.8) in this case. From equation (B.8), and the definition of WPT in equation (B.6), it is clear that the firm's net revenue can no longer be written as a function of $P(1-t_s)$ and other variables. As a consequence the simple equivalence between prices and severance tax rates no longer holds. Because the WPT is applied only to the value of production that exceeds its value at base period prices, a given price reduction causes a more than proportionate reduction in the firm's WPT liability. An implication of this is that the sensitivity of the firm's net revenues to changes in p is reduced.

Despite this the effects of a severance tax on exploration and development can still be simulated accurately by applying the simple equivalence rule stated earlier so long as the estimated price responsiveness of exploration and development used in this simulation pertains to the period prior to introduction of the WPT. To see this it is convenient to rewrite equation (B.8) as

$$N = (1-t_f)(1-t_c)(Hq - c), \text{ where} \tag{B.11}$$

$$H = p(1-t_s)(1-r)(1-t_w(1-p_b/p)). \tag{B.12}$$

The firm's exploration and development decisions will in general be functions of H as well as t_f, t_c, c, and other variables. The econometric model we use to analyze exploratory and development drilling is presented in detail in chapter 5. Suppressing nonprice terms and ignoring other details, the model postulates a simple loglinear relationship between H and the level of drilling activity

$$ln(W) = a + bln(H) \tag{B.13}$$

where W represents the number of wells drilled in a given period. Using equation (B.12) to expand H yields

$$ln(W) = a + bln(1-r) + bln(1-t_s)$$

$$+ bln(p) + bln(1-t_w(1-p_b/p)). \tag{B.14}$$

During our sample period 1948–81, r and T_s were approximately constant and were therefore incorporated into the constant term in the regression. The WPT rate was of course zero prior to 1979. For simplicity, therefore, the term in $[1-t_w(1-p_b/p)]$ was proxied by a dummy variable that took the value zero prior to 1979 and unity thereafter. Within this specification the price coefficient b is the appropriate elasticity to use when simulating the effect of a new severance tax on drilling activity.

There remains the effect of the WPT on the decision of when to shut in a producing well. Our model of the shut-in decision specifies that a well will continue to operate so long as $H>c/q$. As explained in chapter 5 and appendix C, our estimates of the production effects of changes in the shut-in date require information on the effect of a new severance tax on *H*. From equation (B.12), however, it is clear that an *x*-percent severance tax will reduce *H* by exactly *x* percent, so that simulation of the shut-in effect is straightforward.

It should also be noted that the WPT contains a provision that limits the base of the tax to 90 percent of the net revenue attributable to production from a given field. When this occurs the firm's net revenue from production becomes

$$N^1 = (1-t_f)(1-t_c)(1-.9t_w)(Hq-c), \text{ where} \tag{B.15}$$

$$H = p(1-t_s)(1-r). \tag{B.16}$$

When the net income limitation is applicable, the effect of an *x*-percent severance tax is to reduce *H* by *x* percent. Simulation of the shut-in effect remains straightforward, therefore, even if the net income limitation is incorporated into the analysis.

The preceding analysis is applicable to the effects of a severance tax on most properties. These results were, however, based on the assumption that the operator is not liable for severance tax payments on royalty production. Although this appears to be the norm, some leases in Kern County, and perhaps elsewhere, specify that the operator is liable for severance taxes on all production from a property. The effect of such a provision would quite naturally be to multiply the operator's severance tax liability per barrel by $1/(1-r)$ [see equation (B.5)]. Consequently a 6-percent severance tax would, for example, become equivalent to a 7-percent price reduction if the royalty rate were one-seventh. There is no published information on the prevalence of such lease arrangements nor on the exact royalty rates applicable to them. The preceding consideration is therefore ignored in our empirical analysis of supply responses, and this may impart a slight conservative bias to our estimates of the effect of a new severance tax on crude oil supply in California.

Chapter 5

Effects of a Severance Tax
on Crude Oil Production

Estimates of the effect of a severance tax on crude oil production in California are important for several reasons. Most obviously the size and timing of the supply effect will determine the revenue yield of the tax. Less obviously, reductions in petroleum production will, in general, reduce the taxable incomes of producing firms and royalty owners, and thus reduce the yields of individual and corporate income taxes. In addition the tax will affect employment in the extraction sector and supporting industries and the size of the employment impact is directly related to the size of the production impact. Tax-induced reductions in output arise from two sources—premature abandonment of existing oil wells and losses of production from wells not drilled as a result of the tax. These two production effects are examined in successive sections of the present chapter. In the final section of this chapter production results are modified to depict the effect of exempting small producers (less than 100 bbl. per day) from the tax. Throughout this chapter the theory that underlies our estimates is discussed only briefly. For a more detailed explanation see appendix C.

When examining the effects of the proposed severance tax on production, we are attempting to discern the sensitivity of crude oil supply to the (net) price received by producers. Implicit in our analysis, therefore, is a supply model for crude oil produced in California. The general form of this supply model was discussed briefly in our survey of oil and gas supply studies. For those not familiar with the concept of a supply function, or an economic relationship between price and quantity supplied, a brief intuitive discussion may be useful.

The relationship between price and supply is a familiar concept in economics and one that has been studied extensively for a variety of products, including crude oil. The basis for the relationship is the plausi-

69

FIGURE 5.1
1985 Oil Supply, Region 2

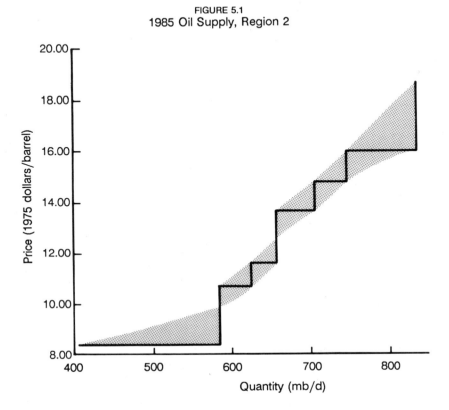

SOURCE: U.S. Energy Information Administration (1978).
NOTE: Region 2 is onshore California.

ble notion that as the price of a good increases, producers are willing to supply more of it. Likewise if the net price producers receive is reduced, as would be the case if a tax were imposed on production, supply falls. Figure 5.1 illustrates the relationship between price and the quantity of crude oil supplied. This schedule was estimated by the U.S. Department of Energy as part of its analysis of price decontrol. The supply curve drawn in figure 5.1 pertains to crude oil produced onshore in California. Thus it is of obvious relevance to our study. We could not, however, rely on it to estimate the production effects of the tax because it does not provide sufficient detail on the timing of production responses. It is very useful, however, in providing an independent check of the results we obtain, as noted later.

In all of the estimates we present, it has been assumed that the severance tax rate would be 6 percent, it would not be applied to lease fuel used in steam generation, and royalty owners (rather than working-interest owners) would be liable for the tax levy on royalty production. It is also assumed that certain fields would be exempt from the tax: federal production from the Elk Hills N.P.R., production from all federal O.C.S. leases, and production from the state-owned Long Beach Tidelands. It is also assumed that real crude oil prices in California will remain constant at current levels. Implicitly, then, the supply of crude oil imports to California is taken to be perfectly elastic, so that none of the tax will be passed forward to refiners or consumers (see chapter 6 for further discussion). There is of course no consensus on the future course of world crude oil prices, and in our analysis crude prices are assumed to remain constant in real terms over the time horizons examined. For the purpose of isolating the impact of a severance tax on petroleum production, this is an appropriate benchmark assumption. While rising or falling real crude oil prices would alter the impact of a severance tax on production, the preparation of price forecasts was outside the scope of the present study.

The crude oil supply model developed for California has two basic components. The first component specifies the relationship between price and rates of production from existing oil wells. The time profile of production from a developed oil field is assumed to be fixed by geologic factors. Over time, as reserves in the field are depleted, the rate of production declines. The operator influences production from an existing field only by deciding to shut in wells producing from it. Producing fields are shut in only when net revenues from production fail to cover operating costs. The second component of the model is an econometric analysis of exploratory and development drilling. This component is similar to econometric models that have appeared elsewhere in the literature in that drilling is specified to respond primarily to crude oil price. Other determinants of drilling activity are a pure trend, the observed trend in the API gravity of new production in California, and introduction of the WPT. Given an estimated model of the response of drilling and reserve additions to price, the effect of price on production from future wells may be modeled by reapplying the first component of the model.

5.1 Crude Oil Supply from Existing Wells

By reducing the net price received by producers, the severance tax would shorten the economic life of existing wells. Our estimate of the quantitative importance of this effect is based on actual economic and geologic data pertaining to each of California's 245 onshore (and offshore within the 3-mile limit) oil fields. Our chief source of information on existing fields is the series of annual reports from the California Division of Oil and Gas (1981) (hereafter referred to as DOG). These extensive reports contain information on annual production, proved reserves, number of producing and shut-in wells, average daily production per well, and method of secondary or tertiary recovery (if applicable) for every California oil field. The information is based on a census of all production and embodies the best information available to the DOG. These reports contain data required to estimate decline rates, pre- and post-tax shutdown dates, and the total tax-induced loss in production for each field. There is of course some uncertainty in the estimates for any particular field, due to the inherent uncertainties in estimating proved reserves. Because such a large number of fields are covered, however, aggregate estimates of output losses are considerably more reliable than estimates for any individual field.

We begin by outlining the way the economic lifetime of a well (or field) is determined and how it is influenced by a severance tax. We then describe how the parameters needed to derive empirical estimates of output losses are obtained. Our model of well lifetime and production is quite standard (see, for example, Kalter et al. 1975; and Camm et al. 1982).

5.1.1 Supply from Existing Wells: Theory

Typically, annual production from an individual oil well declines monotonically over its producing life. Unless new investment, for example for enhanced recovery, is undertaken, the annual percentage decline is largely determined by geologic conditions. Because production costs per well are not highly sensitive to rates of output, it follows that net cash flow from a well will also decline as the well ages. The rate at which cash flow declines each year depends both on the reduction in output and the rate of crude oil price change (if any).

The rational operator ceases production when net cash flow falls to zero. The overall process, and the characterization of the economic limit, is depicted in figure 5.2. The downward sloping curve shows annual production from a representative well as a function of the well's age. Break-even production occurs when the revenue obtained from oil produced is just sufficient to cover operating costs per period. This break-even production, in the absence of a severance tax, is labeled $Q(T_0)$.

FIGURE 5.2
Effect of Tax on Production from Existing Wells

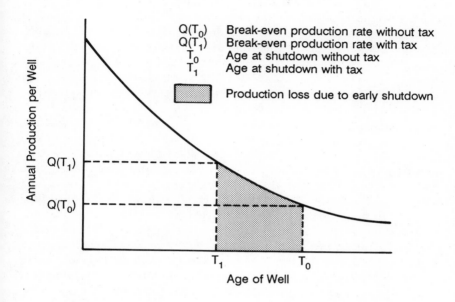

Imposing a severance tax is equivalent to increasing the operator's costs (or decreasing in revenue per barrel). A severance tax increases the amount of output required for the operator to break even, as shown by $Q(T_1)$ in figure 5.2.

Note that when the break-even production level is increased, the economic life of the well is shortened from T_0 to T_1. This is the premature abandonment effect. The size of this effect depends on the size of the tax, the original life of the well (T_0), and the annual production decline rate. Associated with this premature abandonment decision is a loss in total production from the well. Recalling that the height of the curve in figure 5.2 represents annual production, total output from the well is represented by the area under this curve, from the horizontal axis to the well's economic limit (the age at which breakeven production occurs). For the situation depicted in figure 5.2, the tax reduces the well's producing life from T_0 to T_1 and causes a loss in total output that is represented by the shaded area. This is the production effect associated with premature shut-in.

The relevant price in the operator's abandonment decision is the net price received after all taxes, royalties, and other deductions from revenues have been subtracted. In specifying this net price and the way in which it would be affected by a new severance tax, we assume that (1) severance tax payments would be deductible from the WPT; (2) severance taxes would be deductible from federal and state corporate income taxes; and (3) royalty owners, rather than working-interest owners, bear the liability for severance taxes on royalty production. These assumptions all reflect typical tax and operating institutions in the industry. (See appendix *B* for further discussion.)

The formal model used to estimate the size of the shut-in effect follows the same logic. The formal details of the approach appear in appendix *C*. In brief, each well is assumed to be characterized by a fixed production decline rate, and production from the well continues until the net cash flow from the well is zero. It should be noted that the assumption of a fixed decline rate may lead to some underestimation of the effects of the tax. We do not model the changeover of some wells from primary to enhanced recovery or from secondary to tertiary recovery in the period after imposition of the severance tax. Such a changeover would have the effect of increasing total production from the well over its economic lifetime. Since the tax makes production less profitable, some wells that would otherwise have been modified for enhanced recovery will not be; hence the extra output they would have produced will be lost.

5.1.2 Supply for Existing Wells: Estimation

The DOG *Annual Reports* contain all the information required to implement the model just described. It is their estimate of proved reserves for each field, which includes information on prices and costs, and allows us to determine the pretax life of wells. (One of the primary functions of DOG is to maintain up-to-date estimates of proved reserves, and their recent *Annual Reports* show that these estimates are indeed updated frequently.) These data constitute the most reliable published source of the information on California oil reserves presently available.

The primary empirical task required to implement this approach is estimating production decline rates for each tax eligible field in the state. The details of our approach are presented in appendix *C*. To summarize, we began by assuming that the shut-in effect for a given field can be adequately captured by examining only a single average or representative oil well in that field. That is, we ignore variations in the ages, decline rates, and so forth, among wells in a given field. In actual estimation of

TABLE 5.1
Impact of a 6-Percent Severance Tax on Production from Existing Wells

	First 10 Years	First 20 Years	First 30 Years	Full Lifetime of Existing Wells
Cumulative output (million bbl.)	2,121	3,259	3,744	3,788
Cumulative reduction in output caused by tax (million bbl.)	37	160	391	398
Percent reduction in output	1.8%	4.9%	10.4%	10.5%

decline rates, it was necessary to employ more than one technique. For most fields the decline rate for existing wells was estimated as the average annual decline rate from the year of maximum production during the 1976–81 period to 1981, the most recent year for which data are available. Rather than simply estimating decline between 1976 and 1981, we used the year of maximum production to avoid underestimating the decline rate, which might occur if production were increased by initiating enhanced recovery between 1976 and 1981.

Not all fields' decline rates could be estimated in this manner. Some fields achieved maximum production per well in 1981, so that there was no recent historical record of decline from which to obtain an estimate. For other fields the average decline rate estimated over the 1976 to 1981 period exceeded the maximum possible decline rate implied by current reserves and production estimates (see appendix C). In both of these cases the decline rate was estimated statistically. Details of this procedure are presented in appendix C.

Equipped with estimates of per-well decline rates by field, computation of field lifetimes and production impacts for each California oil field and the state as a whole is relatively straightforward. Results for the state are summarized in table 5.1. Individual estimates for each field in the state are presented in appendix C.

Table 5.1 shows cumulative effects for the first 10, 20, and 30 years after imposition of a 6-percent severance tax and for the full lifetime of all existing fields. Examination of these findings shows that the impact of the severance tax on production from existing wells accelerates over time, a result that follows from the age distribution of producing wells in California. Since the tax affects production from existing wells by moving the shutdown date closer to the present, it follows that the impact on *existing* wells will be concentrated at around the time that currently producing wells would have shut down without the tax. The average pretax remaining lifetime of existing California oil wells is 23.6 years, where well lifetimes are weighted by reserves. The weighted average post-tax expected remaining lifetime is 20.7 years, nearly 3 years less. Thus most of the production impact from existing wells occurs during the latter half of the 30-year interval following imposition of the tax. The build-up of production losses from tax-induced shut-ins is even more apparent from figure 5.3. Here the cumulative production effect from existing wells is plotted over the entire 30-year period. During the entire lifetime of existing wells, the tax would reduce production by more than 10 percent.

5.2 Drilling, Development, and Crude Oil Supply from Future Wells

The other major effect of the severance tax on production arises from the fact that the tax would deter drilling of new oil wells, both for exploration and development. Tax-induced reductions in drilling activity reduce the amount of new oil found (exploratory drilling) and lower the incentive to search for extensions of known fields and pools (development drilling). The first step in quantifying the magnitude of these impacts is to determine how drilling activity responds to the price of crude received by producers. Here we have constructed separate, though similar, models for exploratory and development wells.

Our approach to this problem is based on statistical analysis of the historic relationship between (inflation-adjusted) crude oil prices and drilling activity. Because the details of our econometric model are presented in appendix *C,* the discussion here is brief and informal. Our statistical model proceeds from the assumption that decisions to drill both exploratory and development wells are based on expectations of future prices, and that these expectations adjust over time in response to observed price changes and other exogenous factors. In this fashion our framework allows the drilling response to price changes to be phased in and estimates the speed at which this response proceeds. The appropriate price in this model is again the net price received by producers. Because only an index of net price is needed for statistical analysis, we use the gross wellhead price (deflated by the producers' price index) as a proxy for net price. A more detailed explanation of this procedure is presented in appendix *B.*

Drilling costs were explicitly excluded from our model. In the long run there is little reason to believe that the costs of drilling equipment are changing relative to other manufactured goods. In the short run drilling

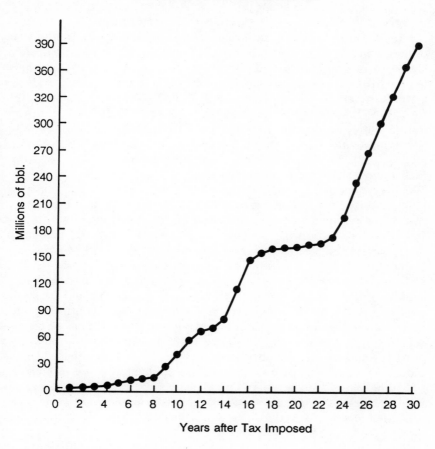

FIGURE 5.3
Cumulative Loss in Production from Existing Wells
Due to 6-Percent Severance Tax

costs can rise and fall radically when the demand for drilling equipment shifts, and these short-run cost variations serve to equilibrate the supply and demand of drilling equipment. In effect, however, short-run drilling costs are determined by crude oil prices. If short run drilling costs were "held constant" in our model by including a cost index in the regression equation, the simulated response of drilling to price would be overstated. By excluding drilling cost our price term is allowed to capture any impact that crude oil price exerts on such costs in the short run.

In addition to expected net price we also include a pure trend in our drilling equations. This trend captures such factors as the diminishing return to exploration and development that occurs because the most likely prospects are explored and exploited first. For the United States as a whole, reserves added per new well drilled have been declining over time (U.S. Energy Information Administration 1979, pp. 40–41); in California initial production per successful new well has fallen recently according to the Conservation Committee of California Oil Producers (1982, p. 92). We therefore expect a negative trend in both the exploratory and development drilling equations.

One exogenous variable that we include in our expected price equation is the ratio of the number of new wells producing light oil to the number of new wells producing heavy oil. Since World War II the gravity mix of California production has swung radically. In 1949 fully 80 percent of newly completed wells produced light (above 20 degrees gravity) crude. By contrast only 18 percent of new wells produced light crude in 1982, and considerable fluctuations in the gravity ratio have occurred from year to year. The ratio of light to heavy producers should enter the expected price equation because our California crude oil price variable does not control for changes in average gravity. Thus much of the observed price movement since the late 1940s actually reflects changes in gravity distributions. By correcting for gravity changes in the drilling model, we allow our price variable to reflect changes in the profitability of finding and producing a barrel of oil of given gravity. (This variable was not statistically significant in our estimated equation for exploratory wells and hence was excluded from our final specification of that equation.)

Finally, included in our price expectations equation is a dummy variable marking the period during which the WPT has been in effect. Since the WPT is effectively an excise tax, it is clearly an important determinant of net price. Moreover, as explained in chapter 4 and in order to use the estimated price response from the model to simulate the introduction of a new severance tax, it is necessary to control for the effect that the

WPT has on the price responsiveness of drilling. The coefficient of the WPT variable cannot be estimated with much precision, however, since the tax has been in effect only since 1980.

Estimates of the drilling model for exploratory and development wells are presented in appendix *C*. Of particular interest are the price and lagged-adjustment terms in the two equations. In both cases the price coefficients are positive and statistically significant. (The probability that the positive relationship between price and drilling is due to mere chance is less than 1 percent for exploratory holes and less than 6 percent for development wells.) In both cases the short-run price elasticity is about 0.42–0.43. Computation of long-run elasticities is complicated by the inclusion of the trend term. However, as our simulation results (presented later) show, the long-run sensitivity of drilling to price implies elasticities in the range of 1.1–1.3.

To simulate the effects of a new severance tax, the model was used to project new drilling over the next 30 years. In baseline simulations the real net price of crude oil was assumed to remain constant. To examine the effects of the proposed tax, drilling was simulated with the price term reduced by 6 percent. Our simulations also reflect current law requiring the WPT to begin to be phased out no later than January 1991. Specifically, our model assumes that the WPT lapses in 1992. As the following results indicate, this should cause a significant though temporary increase in future drilling activity.

Forecasts of drilling activity over a 30-year period after imposition of the tax are shown in figures 5.4–5.6. Both exploratory drilling and completion of new producing wells show an initial increase followed by a decline, then another jump when the WPT is phased out, again followed by a decline. The initial increase indicates that drilling in California, though increasing over the past few years, has apparently not yet completely caught up to the real crude price increases of 1979–82. Because some completed producers are drilled as exploratory wells, there is some overlap in the series in figures 5.4 and 5.5. To eliminate this it was assumed that the fraction of exploratory wells that become producers will remain at its average 1976–81 level (26 percent). Figure 5.6 shows total drilling (with this overlap eliminated) both with and without the tax. Overall the tax is estimated to reduce total drilling by 6 to 7 percent in the long run. Moreover, the transition to this long-run response is very rapid. This is evident from the plots in figures 5.4–5.6; for total drilling the industry's response to the tax (as measured by the vertical distance between the plots) is virtually complete by the third year.

Our projections of future drilling activity are translated into future

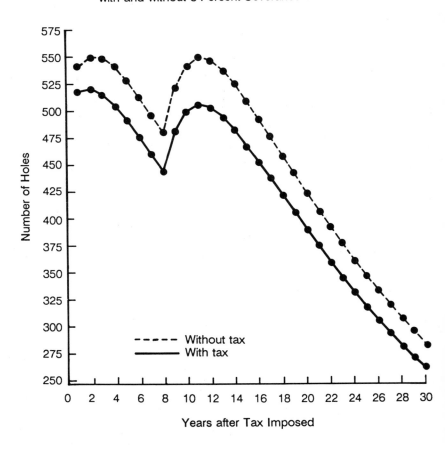

FIGURE 5.4
Exploratory Holes Drilled per Year,
with and without 6-Percent Severance Tax

Number of Holes

Years after Tax Imposed

- - - - - Without tax
——— With tax

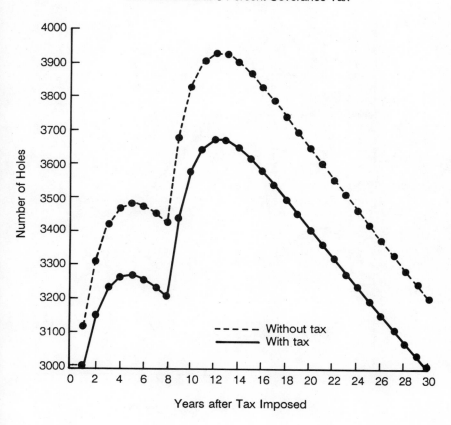

FIGURE 5.5
Completed Producing Wells Drilled per Year,
with and without 6-Percent Severance Tax

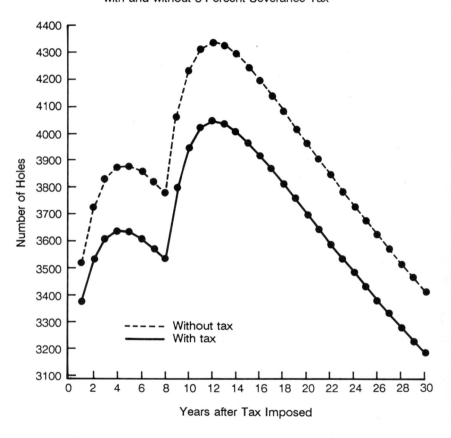

FIGURE 5.6
Total New Wells Drilled per Year,
with and without 6-Percent Severance Tax

Effects of a Severance Tax on Crude Oil

TABLE 5.2
Impact of a 6-Percent Severance Tax on Production from New Wells

	First 10 Years	First 20 Years	First 30 Years
Cumulative output without tax (mmbbl.)	1,525	4,195	6,631
Cumulative reduction in output caused by tax (mmbbl.)	85	252	408
Percent reduction in output	5.6%	6.0%	6.2%

production effects by using the model of supply from existing wells just described. To carry out this analysis two additional pieces of information are needed: the initial production rate from newly completed wells and the annual production decline rate of new wells. Both of these questions were examined statistically using data from our sample of California oil fields. The details of our methodology and results appear in appendix C. To summarize, our estimates of initial production from new wells account for the decline in new-well productivity that has been evident during the last 25 years. This negative trend is another reflection of the industry's natural tendency to exploit the most promising drilling sites first. Regarding the annual decline rate for new wells, a range of estimates was considered. The end points of this range correspond to (1) the simple unweighted average decline rate for fields in our sample and (2) the weighted average decline rate, where individual field decline rates were weighted by reserves. The two estimates are approximately 9 percent and 3 percent, respectively. Accordingly two sets of new-well production impacts were computed.

If a 9-percent average annual decline rate for new wells is assumed, the percentage cumulative reduction in total output from new wells ranges from about 4.1 percent initially to about 6.2 percent at the end of 30 years. Cumulative output reductions for 10-, 20-, and 30-year periods are presented in table 5.2. (The long-run price elasticity of new development wells was 1.091, which would imply a long-run annual output reduction of 6.5 percent associated with a 6-percent severance tax.) The percentage reductions in cumulative output each year are the same for the 3-percent decline rate, since the percentage reduction for any given vintage of new wells depends only on the percentage reduction in drilling that year and not on the decline rate. Of course the *total* output, and reduction in output from new wells, will be larger with the lower decline rate.

TABLE 5.3
Impact of a 6-Percent Severance Tax on Total Production

	First 10 Years	First 20 Years	First 30 Years
Cumulative output without tax (mmbbl.)	3,646	7,454	10,375
Cumulative reduction in output caused by tax (mmbbl.)	123	412	799
Percent reduction in output	3.4%	5.5%	7.7%

5.3 The Impact of a 6-Percent Severance Tax on Total Production

The aggregate effect of the severance tax on output is the sum of the effects from premature shut-in of existing wells and the loss of potential output from new wells not drilled. Figures for 10- , 20- , and 30-year output effects appear in table 5.3. Over time, production from new wells becomes a progressively larger fraction of total cumulated output. Assuming a 9-percent decline rate for new wells, the fraction of cumulative output due to new wells rises from 42 percent after 10 years to 64 percent after 30 years.

Production effects from existing fields and new wells are combined in figure 5.7. Estimated production both with and without the tax is larger when a 3-percent decline rate is used, but the time pattern is very similar to that shown in figure 5.7; hence these estimates are not reported separately here. These annual losses are cumulated in figure 5.8. In figure 5.9 cumulative production losses attributable to the tax are expressed as a percent of projected cumulative output without the tax. The actual yearly output projections used in these plots are presented in appendix *C* (see tables C.4 and C.5).

For a measure of the sensitivity of production to the proposed tax, we note that the tax is expected to reduce cumulative production by 3.4 percent over the first 10 years it is in force, by 5.5 percent over a 20-year horizon, and by 7.7 percent over a 30-year time span. When compared to the 6-percent reduction in net price that the tax causes, these production responses imply price elasticities of supply of 0.57, 0.92, and 1.28, respectively, for these three time periods. These elasticity estimates permit our results to be compared to those obtained independently by other researchers. The summary of crude oil supply elasticity estimates introduced in chapter 4 is reproduced here for convenience as table 5.4. Overall our estimates are within the general range of those obtained in other studies. Note in particular the close correspondence between our

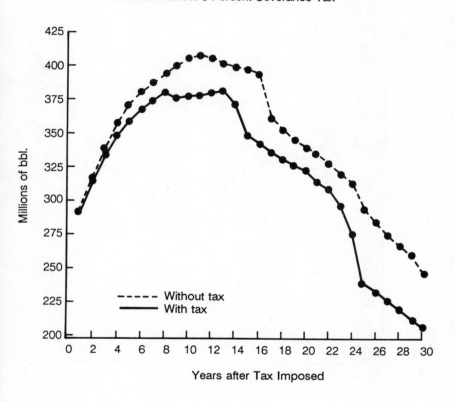

FIGURE 5.7
Annual Production from New and Existing Wells,
with and without 6-Percent Severance Tax

Years after Tax Imposed

Millions of bbl.

- - - - Without tax
——— With tax

85

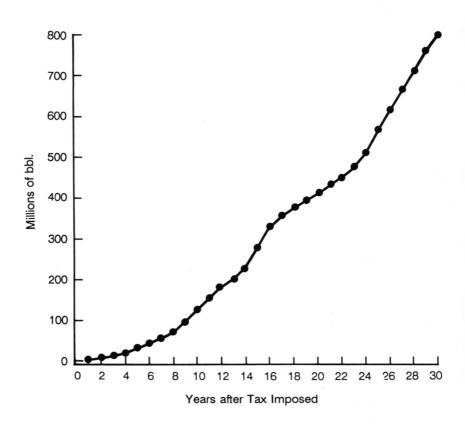

FIGURE 5.8
Cumulative Loss in Production from All Wells
Due to 6-Percent Severance Tax

86

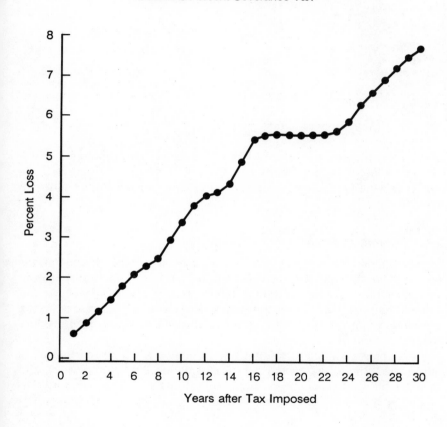

FIGURE 5.9
Percentage Loss in Output from All Wells
Due to 6-Percent Severance Tax

87

TABLE 5.4
A Summary of Crude Oil Supply Elasticity Estimates

Author or Institution	Elasticity Estimate	Description
U.S. Geologic Survey (1980)	1.14 1.0	For $22–$29 price range For $29–$37 price range
Mancke (1970)	1.0–2.0	
Erickson and Spann (1971)	0.9	Elasticity of reserves with respect to price
U.S. Cabinet Task Force (1970)	1.0	From survey of oil producers
Phelps and Smith (1977)	1.0	Adopted from other studies
U.S. Energy Information Administration (1978)	0.73–1.25	For $21–$28 price range specific to California onshore production

elasticity estimates and those of the U.S. Energy Information Administration (1978). To our knowledge these are the only other crude oil supply elasticity estimates presently available for onshore California specifically.

5.4 The Effect of Small Producer Exemptions

Some of the recent proposals for a severance tax provide for exemptions for small producers. Two such exemption levels considered as part of severance tax proposals in California have been 36,500 bbl. per year and 100,000 bbl. per year. To examine the effect of these exemptions on the production impact of the tax, we collected information on 1981 production by field and by company.

Each company's taxable output was computed as the proportion of its output over the initial excluded level (36,500 or 100,000 bbl./year), and this taxable output was allocated proportionately across all the company's operations. That is, the taxable proportion of each company's output in each field was set equal to the overall taxable proportion of that company's output. Total taxable production per field was computed as the sum of the taxable outputs of each of the companies operating in the field. The model of premature shut-in was then applied to this fraction of each field's production.

This procedure resulted in 48 fields being excluded from taxation entirely under the 36,500-bbl./year exemption. The proportion of taxable production from the other fields varied according to the size of the operators. The fields most affected by exclusions, however, were typically very small. This is evident from a comparison of the tax-induced reduction in total output with and without the exclusionary provision. Without exclusions the total reduction in output due to a 6-percent tax, over the full lifetime of existing fields was 397.6 million bbl., or 10.5 percent of total output without the tax. With the 36,500-bbl. exclusion the reduction in output is 384.1 million bbl., which is only 3 percent less than the production impact with no exclusions. Even with the 100,000-bbl./year exclusion, the reduction in total lifetime output from existing fields is 377.1 million bbl., or 95 percent of the reduction without exclusions. It is clear that given the present structure of the crude oil industry in California (that is, the current mix of large and small firms), the exclusions proposed to date would do little to reduce tax-induced losses in output from existing wells.

To calculate the effect of such exclusions on new drilling, it is necessary to specify the proportion of new drilling that would be effectively excluded from the tax. Again assuming the current structure of the industry remains unchanged, we calculated the weighted average of *taxable* production by field (the weights being the 1981 production of each field) and assumed that the same proportion of new drilling would be taxable, that is, would not be excluded. The results of this analysis indicate that such exclusions would do little to mitigate the impact of the proposed tax on drilling and new production. Our estimates for the 36,500bbl./year exemptions indicated that the drilling impact would only be lowered by about 5 percent; the 100,000-bbl./year exclusion would reduce the adverse impact of the tax on drilling by about 7 percent. The effect of these exclusions on production from new wells would, given our assumptions, be of the same proportions.

If any of the proposed exemptions resulted in a radical shift in the composition of the industry in California, then the impact of the tax on production could be altered more substantially. For example, a small producer exemption might induce large firms to sell marginal oil wells to small firms as those wells approached the economic shut-in date under the tax. If such behavior were permitted under the exclusion policy, it would clearly mitigate the premature shut-in effect of the tax. However, we are not aware of any analysis of experience with similar exemptions elsewhere that would allow us to forecast the magnitude of this effect. If such behavior were universal so that all oil were produced by small producers it could eliminate the early shut-in effect of the tax. Such an outcome is extremely unlikely, however. The extent of any tax-induced property transfers would depend on the cost disadvantages of dividing oil-producing properties into small units. It would obviously be necessary to divert resources from other endeavors to negotiate the sale of property. Moreover, some economies associated with large-scale recovery operations would be lost. The latter point is particularly important in situations where secondary or enhanced thermal recovery is practiced. Such methods are of course commonly applied in California as wells approach the end of their economic lives. As currently practiced, steam and water injection are applied to groups of wells rather than to one well at a time. It is not at all clear that such techniques could be applied efficiently on a scale small enough to qualify for either of the exemptions proposed to date. But beyond such qualitative statements, little more can be said at this point about the actual magnitude of the effect of tax-induced changes in the structure of property ownership on premature shut-in.

With regard to the effect of such behavior on new drilling, the impact of proposed exemptions is less uncertain. The drilling decision depends on expected net revenue during the entire life of a well. Transfers of the type just discussed, however, would affect receipts only during the last few years of a well's economic life and thus do little to alter the overall economics of a drilling prospect. The exclusions might encourage a proliferation of small drilling and production firms. At 1982 levels of initial output per well, however, as few as three new wells would put a small producer over the 36,500 bbl./year level. Again, excluding the first 36,500 or 100,000 bbl./year would appear to have very little effect on overall tax-induced reductions in output from new wells.

We have also examined the effect of excluding stripper wells (10 bbl./day or less) from the proposed severance tax. While a large fraction of California wells are in the stripper well category, only 18 percent of 1981 output was produced by strippers, and only 16 percent of reserves is associated with such wells. If all wells producing 10 bbl./day or less in 1981 were exempted from the tax, the estimated total tax-induced reduction in output would be 369.1 million bbl. This figure is 93 percent of the output reduction from existing wells without the stripper exemption. There is, however, an additional complication that renders the ultimate consequence of a stripper exclusion uncertain.

Obviously some existing nonstripper wells will become strippers toward the end of their useful lifetimes and would thus fall under the exemption. Here the stripper well exclusion mitigates premature abandonment. On the other hand some producers would prematurely reduce the output of nonstripper wells to qualify them for the stripper exemption, thus enhancing the production effect. The result of these offsetting forces cannot be calculated with any precision. Our analysis indicates that in the absence of significant changes in the structure of the industry brought about by small producer or stripper exemptions, such exemptions would not significantly alter the production or employment effects of the tax. However, the difficulty of forecasting alterations in the structure of the industry to take advantage of exclusions necessarily renders the final impact of such exclusions uncertain.

Appendix C

Effects of a Severance Tax

on Production:

Analytical Details

This appendix develops the formal methodology used to specify the supply model and project production impacts of the proposed severance tax. It also presents detailed estimates of all relevant parameters. The first two sections of this appendix explain theoretical and empirical aspects of the supply of oil from existing wells and the tendency to abandon wells prematurely when the tax is imposed. The third section is concerned with decisions to drill new wells and the resulting impact on production from newly discovered or developed reserves.

C.1 Supply from Existing Wells: Theory

Our model of the producer's decision to cease operation from a particular well is analytically similar to analysis developed by Kalter et al. (1975) and by Camm et al. (1982). We begin by specifying the age at which cash flow falls to zero. If a is the decline rate, $P(t)$ is the price net of output-related production costs and taxes at time t, q_0 is the output from the well at time zero (the present), and $B(t)$ represents non-output-related costs measured in dollars per unit of initial capacity, then the economic lifetime of the well T_0, is given by solution of the equation

$$P(T_0)q_0 e^{-aT_0} - B(T_0)q_0 = 0. \qquad (C.1)$$

It makes no difference whether $B(t)$ is measured in units of capacity at the time the well was completed or in terms of capacity at time zero. The interval from initial construction to time zero would merely yield a constant that drops out of equation (C.3).

If $P_s(t)$ is the net price with the severance tax, and T_1 is the lifetime of the well given imposition of the tax, then

$$P_s(T_1)q_0 e^{-aT_1} - B(T_1)q_0 = 0 \qquad\qquad (C.2)$$

so that

$$T_1 - T_0 = (1/a)\ln[P_s(T_1)/P(T_0)] \qquad\qquad (C.3)$$

provided $B(t)$ does not change. (The effects of intertemporal changes in B will be shown momentarily). If p is the rate of change of real net oil prices and s is the severance tax rate,

$$P_s(T_1) = (1-s)P(T_0)e^{p(T_1 - T_0)}. \qquad\qquad (C.4)$$

(Note that $T_1 < T_0$.) Thus if s is not too large,

$$T_1 - T_0 = -(s/a) + (p/a)(T_1 - T_0) \qquad\qquad (C.5)$$

or

$$T_1 = T_0 - (s/a)/[1 - (p/a)]. \qquad\qquad (C.6)$$

For the well to have finite life the output decline rate is required to be larger than the expected rate of increase in real net oil prices, or $(p/a)<1$, so the denominator of the second term on the right of equation (C.6) is positive and T_1 is always less than T_0. It is also easy to see that if $B(t)$ changed at a rate b per year, then equation (C.6) would be modified to read

$$T_1 = T_0 - (s/a)/[1 - (p/a) + (b/a)] \qquad\qquad (C.7)$$

The empirical estimates we report are based on the assumption that $b=0$ and $p=0$, so that

$$T_1 = T_0 - (s/a). \qquad\qquad (C.8)$$

Different assumptions about real net oil prices and costs can be incorporated into our framework by appropriate substitution of equation (C.7) for equation (C.8) in the calculations that follow.

Equation (C.8) gives the change during the economic lifetime brought about by the tax; to obtain the effect of the tax on output of the well, it is necessary to determine T_0, the pretax expected lifetime, and calculate the amount of output the well would have produced between T_1 and T_0.

Proved reserves are defined as the amount of oil economically recoverable at current prices and costs. Thus they equal total expected production from a well over its lifetime, assuming p and b equal zero. If W_0 is the number of producing wells in a given field at time zero, R_0 is proved reserves at time zero, and $Q_0 = W_0 q_0$ is the initial output from the field, then the materials balance equation for production from the field is

$$R_0 = \int_0^{T_0} Q_0 e^{-at} \, dt \qquad\qquad (C.9)$$

or

$$R_0 = (Q_0/a)(1-e^{-aT_0}) \qquad\qquad (C.10)$$

which implies

$$T_0 = -(1/a)\ln[1 - (aR_0/Q_0)]. \qquad\qquad (C.11)$$

If the well produced forever, the total oil recovered would be

$$R_\infty = \int_0^\infty Q_0 e^{-at} \, dt = Q_0/a. \qquad\qquad (C.12)$$

Since $R_0 < R_\infty$, it follows that aR_0/Q_0 must be less than unity; hence T_0 is always defined.

Combining equations (C.8) and (C.11) with the production relationship yields an expression for ΔQ, the reduction in output brought about by the tax:

$$\Delta Q = \int_{T_1}^{T_0} Q_0 e^{-at} \, dt$$

$$= (Q_0/a)(e^{-aT_1} - e^{-aT_0}). \qquad\qquad (C.13)$$

The proportional fall in output is then given by

$$(\Delta Q/Q) = (e^{-aT_1} - e^{-aT_0})/(1 - e^{-aT_0}). \qquad\qquad (C.14)$$

Differentiating equations (C.14) and (C.10) suffices to show that this proportional reduction in output is inversely related to the well's decline rate.

$$\frac{d(\Delta Q/Q)}{da} = - \frac{e^s - 1}{(e^{aT_0} - 1)^2} \cdot \frac{e^{2aT_0}R_0}{Q_0} < 0 \qquad (C.15)$$

since $e^s - 1 > 0$ for any positive severance tax rate.

It is also possible to express relationships governing the well's lifetime in terms of the ratio margin P_0/B_0. Since shutdown occurs when

$$P_0 q_0 e^{aT_0} - B_0 q_0 = 0 \qquad (C.16)$$

it follows that

$$P_0/B_0 = 1/[1 - (aR_0/Q_0)] \qquad (C.17)$$

and this ratio margin provides an alternative expression for how close the well is to shutdown. Since

$$P_0/B_0 = e^{aT_0} \qquad (C.18)$$

from equation (C.16) it is easy to see that the closer the well is to shutdown (the smaller is T_0), the closer is P_0/B_0 to unity for any given value of the decline rate.

It is also a simple matter to compute the production effect for a given well or field over a particular interval of time, say, T_f years. If the post-tax lifetime of the field T_1 is greater than or equal to T_f, there is no production effect during the first T_f years. If $T_1 < T_f$, then production in the first T_f years is

$$Q_f = \int_0^{T_i} Q_0 e^{-at} \, dt \qquad (C.19)$$

where

$$T_i = min\ (T_0,\ T_f) \qquad (C.20)$$

Thus the production effect Q_f is given by

$$\Delta Q_f = 0 \qquad\qquad if \ T_1 \geq T_f$$

$$\text{(C.21)}$$

$$= \int_{T_1}^{T_i} Q_0 e^{-at} \ dt \qquad if \ T_1 < T_f$$

and the proportional production effect $\Delta Q_f / Q_f$ may be computed accordingly.

C.2 Supply from Existing Wells: Estimation

All the information required to implement the model just developed is contained in the *Annual Reports* of the California Division of Oil and Gas. The key equations needed to compute production effects—(C.8), (C.11), (C.13), (C.14), (C.19), and (C.21)—do not explicitly involve prices or costs. Rather the DOG estimates of proven reserves embody all of the information on prices and costs needed to establish the economic lifetimes of wells.

The DOG *Annual Reports* contain data on Q_0 and W_0 for each California oil field; R_0 is given for each field with proved reserves greater than 100,000 bbl. (The method used to estimate R_0 for small fields under 100,000 bbl. proven reserves will be described subsequently.) We examined only onshore fields and those offshore fields within the three-mile limit that would be subject to state taxation. This amounted to a total of 245 fields potentially subject to the tax, of which three—Elk Hills, Wilmington, and Wilmington Offshore—would likely be exempt for other reasons.

Given R_0, Q_0, and W_0, the only additional information required to determine the shut-in effect of the tax is an estimate of the per-well decline rate a. Production per well in a given field in year t is given by Q_t/W_t, so output per well in t_1 and t_2 are related by

$$Q_{t_2}/W_{t_2} = (Q_{t_1}/W_{t_1}) e^{-a(t_2 - t_1)}. \qquad \text{(C.22)}$$

If the interval from t_1 to t_2 is exactly one year, then

$$a = \ln(Q_{t_1}/W_{t_1}) - \ln(Q_{t_2}/W_{t_2}). \qquad \text{(C.23)}$$

It is obvious that the *average* decline rate over an interval of n years will simply be $(1/n)$ times the logarithmic difference of the first and last years' values of (Q/W), since all intermediate terms sum to zero.

Appendix C

As noted in chapter 5, the average annual decline rate in each field examined was estimated from the year of maximum production during the 1976–81 period relative to 1981, the most recent year for which data are available. We averaged between the year of maximum production and 1981 rather than simply comparing 1976 and 1981, in order to avoid underestimating the decline rate as might arise if production increased, for example, following initiation of an enhanced recovery project, during the 1976–81 period.

Some California fields achieved maximum production per well in 1981, so there was no recent historical record of decline to use as a basis for estimating a. For a few cases the average decline rate estimated over the 1976–81 period exceeded theoretical limits. Recall from equation (C.12) that a well producing forever and having decline rate a would produce Q_0/a bbs. of oil. Thus a_{max}, the maximum decline rate consistent with given values of R_0 and Q_0, is Q_0/R_0. An average 1976–81 decline rate greater than a_{max} is therefore not consistent with the basic Q_0 and R_0 data for the field.

In both of such cases, the decline rate a was estimated in a different way. First, a linear regression relationship between a and a_{max} was estimated for those fields for which $a < a_{max}$. This regression was

$$a = .0073 + .4768a_{max} \qquad N = 104 \quad R^2 = .3872 \qquad (C.24)$$

$$(.8076)(8.0282) \qquad (t\text{-}ratios \ in \ parentheses)$$

The a value for cases where the average 1976–81 decline rate was greater than a_{max} was taken to be the value fitted from equation (C.24).

The other major exception to the procedure just outlined was the 57 small fields having reserves less than 100,000 bbl. In these cases reserves were estimated from the relationship between a and a_{max} for the larger fields. An a_{max} was fitted based on the regression of a_{max} on a:

$$a_{max} = .0776 + .8090a \qquad N = 104 \quad R^2 = .3872 \qquad (C.25)$$

$$(8.7123)(8.0292) \qquad (t\text{-}ratios \ in \ parentheses)$$

Reserves were estimated as 1981 production divided by the fitted value of a_{max}. In 6 of the 57 cases where reserves were estimated in this manner, estimated reserves exceeded 100,000 bbl.; in those cases reserves were set equal to 100,000 bbl.

There remained a handful of cases that did not fit any of the preceding categories. For example, in three cases there were no reserves given and output per well was at a maximum in 1981. In those cases the per-well decline rate was set at the state average and a_{max} was fitted from equation (C.25). In eight cases the decline rate was so large that a_{max} fitted from equation (C.25) was still smaller than a. [This occurs if $a > 0.07757/(1 - 0.8090) = 0.4061$.] In those cases, a was estimated from equation (C.24) using the fitted a_{max}. In one case the a_{max} value was so small that the a fitted from equation (C.24) was larger than a_{max}. No further correction was made for this field. All the fields that were exceptional in these ways were very small relative to total taxable production and reserves.

The use of regression equations to estimate a and a_{max} (and reserves) is consistent with our effort to be as conservative as possible in making any adjustments to the underlying data. Prior to fitting any of the a_{max} or a values, the unweighted average a_{max} was 1.84 times the average a. Values of a could have been fitted using this ratio, but the effect would have been a great deal of variation in the resulting estimates of a. (The variance of the fitted values would have been 3.39 times the variance of the a_{max} terms.) Use of linear regression, however, anchors fitted values to the statewide average, using only the additional signal contained in a_{max} to improve the estimate of a. Similarly estimates of a_{max} will show the same sort of regression to the mean. Thus our regression procedure avoids extreme estimates of a and a_{max} that could distort the final results.

The state average of the per-well decline rates varies depending on whether the average is unweighted across fields or weighted by reserves or production. The unweighted average (representing the decline rate typical of a *field* selected at random) is 9 percent per annum. If weighted by 1981 production the average decline rate is 4.3 percent, which is the decline rate associated with a typical *barrel* of oil selected at random from those produced in 1981. If weighted by reserves, the average per well decline rate is only 3.3 percent per year.

Given our estimates of per-well decline rates for each field, effects of the tax on field lifetimes and cumulative production were estimated. Estimates for individual oil fields in California are presented in table C.1.

C.3 Loss of Production due to Wells Not Drilled

To obtain estimates of production loss from new wells, we began by examining the sensitivity of drilling decisions to crude oil price. Our approach is econometric, and we model drilling activity in the standard Nerlove production response framework. Current drilling activity depends on price expectations that are formed adaptively; the current realization of the price embodies the most up-to-date information on future price prospects. Formally

$$W_{it} = c_{i0} + c_{i1} P_{e,t+1} + c_{i2}t \qquad (C.26)$$

where W_{it} = wells of type i (exploratory holes or development wells) drilled in year t, $P_{e,t+1}$ = the real net oil price expected to hold during the period of operation of the wells, and t is a trend. Expectations evolve according to the standard adaptive scheme

$$P_{e,t+1} = P_{et} + g_i(P_t - P_{et}) + h_i X_t \qquad (C.27)$$

where P_t is the current real net price and X_t is a vector of exogenous variables influencing expected real net price. The Koyck transformation yields the estimating equation

$$W_{it} = b_{i0} + b_{i1}P_t + b_{i2} W_{i,t-1} \qquad (C.28)$$
$$+ b_{i3}t + b_{i4}X_t + u_{it}$$

with u_{it} a random disturbance of the usual type. There is no feedback from current drilling to the current price because of our assumption that California crude price is set exogenously by market conditions outside the state.

Readily available data enable econometric estimation of (C.28). The Conservation Committee of California Oil Producers (1982) publishes annual time series data on the number of exploratory holes drilled and the number of producing oil wells completed in California. The American Petroleum Institute (API) (1982) publishes annual average wellhead value per barrel of crude oil by state. The published API price series extend only through 1980, but the 1981 California crude wellhead price is given by the Independent Petroleum Association of America (1982). Nominal crude wellhead prices were converted to real prices using the wholesale price index. We use the gross wellhead crude price as a proxy for the net price received by producers. (See appendix B for further discussion.) Since short-run fluctuations in costs of specialized oil-drilling factors are closely tied to the product price, no drilling cost measure was included. (This point is discussed further in chapter 5.)

As explained in chapter 5, three variables other than price and lagged drilling enter our final estimating equations: a pure trend, the ratio of light to heavy oil-producing wells completed, and a dummy variable that

TABLE C.1
Tax Effects by Field, Production, 6-Percent Severance Tax

Field	Q81	CUMQ	DELTAQ	T0	T1	PTF
Aliso Canyon	500400	5716000	275726	16.7756	15.5556	0.048237
Alondra	1601	8665	211	9.5305	9.0777	0.024390
Ant Hill	112299	1753000	5005	50.9621	49.9823	0.002855
Antelope Hills	131777	976000	136463	8.8452	7.3960	0.139819
Antelope Hills, North	23511	193000	7488	12.7332	11.9316	0.038798
Arroyo Grande	514755	3132000	471393	7.1915	5.9379	0.150508
Asphalto	284300	872000	124071	3.6539	3.0464	0.142283
Bandini	34576	336000	17044	14.0998	13.0384	0.050725
Bardsdale	95927	757000	20332	13.5219	12.8428	0.026859
Barham Ranch	9208	67972	3753	10.4995	9.6611	0.055210
Beer Nose	196	272	18	1.9084	1.7377	0.064688
Belgian Anticline	291170	3111000	153375	15.6087	14.4565	0.049301
Bellevue	140062	1333000	5108	28.7167	28.1103	0.003832
Bellevue, west	188869	3339000	133502	27.2105	25.4640	0.039983
Belridge, north	775692	6748000	352682	12.5329	11.5698	0.052265
Belridge, south	27891990	442358000	18690768	24.0691	22.4673	0.042253
Beverly Hills	2683979	31030000	1490697	16.9956	15.7630	0.048041
Big Mountain	39262	388000	14506	15.4776	14.5261	0.037387
Bitterwater	4480	16478	134	8.9477	8.6979	0.008161
Blackwells Corner	7871	52341	2171	10.1391	9.4724	0.041485
Brea-Olinda	2931422	92000000	2356599	54.5000	51.8369	0.025615
Brentwood	152289	1039000	85078	8.9206	7.9692	0.081884
Buena Vista	2222939	22946000	358916	20.6942	19.9182	0.015642
Burrel	24110	275000	130817	12.1172	6.1682	0.475699

Burrel, southeast	6146	41081	1728	10.1561	9.4823	0.042056
Cal Canal	174965	894000	51893	7.1846	6.5902	0.058046
Calders Corner	3464	34281	5260	11.6662	9.5989	0.153448
Canada Larga	1359	7782	219	9.6823	9.1822	0.028168
Canal	30483	168000	9638	7.7700	7.1326	0.057370
Canfield Ranch	470296	3898000	108639	14.0559	13.3345	0.027870
Cantua Nueva	23821	100000	10183	5.2725	4.6058	0.101831
Capitan	72122	1034000	45736	21.5091	20.0336	0.044232
Capitola Park	1636	10879	451	10.1391	9.4724	0.041485
Careaga Canyon	4782	16119	95	9.0111	8.7896	0.005902
Carneros Creek	13781	127000	6538	13.3243	12.3111	0.051479
Cascade	58621	305000	24696	6.8179	6.0969	0.080970
Casmalia	613725	13651000	473704	35.5293	33.4458	0.034701
Castaic Hills	40028	331000	31191	10.5299	9.2777	0.094233
Castaic Junction	196846	1091000	28070	9.6126	9.1417	0.025729
Cat Canyon	5311495	75963000	3363542	21.4506	19.9781	0.044279
Chaffee Canyon	38261	165000	9803	6.0320	5.5247	0.059410
Cheviot Hills	107828	4635000	75477	85.1534	81.8951	0.016284
Chino-Martinez	13273	34689	32	11.2235	11.0643	0.000911
Chino-Soquel	2974	23465	1575	10.7452	9.7581	0.067100
Cienaga Canyon	22318	176000	9420	11.2977	10.4150	0.053525
Coalinga	8464733	102092000	4832510	17.7942	16.5166	0.047335
Coalinga, east, extension	2001216	7758000	931055	4.7378	4.0538	0.120012
Coles Levee, north	1378240	5667000	93115	8.1237	7.8115	0.016431
Coles Levee, south	393873	5245000	239177	19.8276	18.4394	0.045601
Comanche Point	4288	23291	573	9.5397	9.0841	0.024614
Coyote, east	740385	6219000	327902	12.0762	11.1425	0.052726
Coyote, west	1135044	6795000	131620	11.2681	10.7964	0.019370
Cuyama, south	591439	1862000	187466	3.9619	3.4655	0.100680

Field	Q81	CUMQ	DELTAQ	T0	T1	PTF
Cymric	4292797	47783000	15281003	12.1438	8.0219	0.319800
Deer Creek	75894	1807000	59653	38.5440	36.3528	0.033012
Deer Creek, north	2501	10336	122	9.0126	8.7173	0.01180
Del Valle	143294	1734000	81981	17.8586	16.5774	0.04728
Devils Den	74220	1075000	159618	17.1513	14.1956	0.14848
Dominguez	698227	8879000	47764	34.9069	34.0776	0.00538
Eagle Rest	2911	—	0	—	—	—
Edison	1680313	39464000	1316360	37.9145	35.7452	0.03336
Edison, northeast	6289	103000	4284	24.9543	23.3107	0.04160
El Rio	6111	37765	1293	9.9059	9.3297	0.03425
El Segundo	38117	252000	35877	7.8751	6.5652	0.14237
Elk Hills	63169629	911231000	65994350	19.3322	17.4530	0.07242
English Colony	11572	131000	12278	14.4296	12.7209	0.09372
Esperanza	18989	485000	14386	42.5717	40.3041	0.02966
Eureka Canyon	21051	179000	9409	12.2337	11.2898	0.05257
Fillmore	71574	100000	42993	1,4931	0.8265	0.42993
Five Points	1012	3251	16	9.1101	8.9024	0.00479
Four Deer	8608	57992	2491	10.1823	9.4973	0.04295
Fruitvale	1057542	11749000	859900	14.8567	13.4012	0.07319
Goosloo	12094	103000	3273	13.9297	13.1561	0.03178
Greeley	254236	1498000	29822	11.0020	10.5347	0.01991
Guadalupe	1225194	12683000	631446	15.0871	13.9659	0.04979
Guijarral Hills	36014	189000	8169	7.9190	7.3841	0.04322
Half Moon Bay	945	1296	84	1.8818	1.7135	0.06472
Hasley Canyon	78403	465000	26352	8.3851	7.7031	0.05667
Helm	150346	517000	153007	3.7753	2.5815	0.29595
Holser	45535	292000	16317	9.0955	8.3631	0.05588

Honor Rancho	272366	1411000	81733	7.2877	6.6857	0.05793
Hopper Canyon	50795	437000	65530	10.1737	8.4058	0.14995
Howard Townsite	50802	631000	6809	27.8143	26.9390	0.01079
Huntington Beach	2299224	24400000	609400	18.5614	17.6674	0.02498
Hyperion	9356	95041	16545	11.7803	9.4550	0.17408
Inglewood	3861230	40662000	11641477	11.5941	8.0368	0.28630
Jacalitos	67549	496000	15566	12.0511	11.3869	0.03138
Jasmin	137626	1277000	65617	13.4216	12.4023	0.05138
Jasmin, west	874	5735	230	10.0952	9.4466	0.04004
Jerry Slough	889	3886	54	9.0856	8.7645	0.01386
Jesus Maria	21719	100000	8739	5.9455	5.2788	0.08739
Kern Bluff	173314	1631000	1631000	9.5544	0.0000	1.00000
Kern Front	2540649	29955000	1429452	17.3635	16.1101	0.04771
Kern River	41740421	1076077000	191673134	29.8119	23.8095	0.17812
Kettleman City	16210	17094	1117	1.4445	1.3144	0.06532
Kettleman Middle Dome	1389	4973	37	8.9546	8.7140	0.00743
Kettleman North Dome	276577	1043000	26658	6.5528	6.2330	0.02556
King City	31612	151000	8574	6.7503	6.2006	0.05678
Kraemer	36998	400000	5620	22.3773	21.5812	0.01405
Kreyenhagen	2604	12068	196	9.1835	8.8323	0.01627
La Honda	13333	120000	6217	12.9935	12.0010	0.05181
Lakeside	2682	15223	419	9.6578	9.1655	0.02754
Las Cienegas	916745	7978000	73137	20.4559	19.8563	0.00917
Las Llajas	2148	11045	236	9.4036	8.9884	0.02138
Las Posas	2978	10620	78	8.9564	8.7171	0.00732
Lawndale	13112	160000	7542	18.0217	16.7315	0.04714
Livermore	39451	292000	15852	10.5687	9.7347	0.05429
Lompoc	195192	2640000	17436	35.0022	34.1040	0.00660
Long Beach	3529565	41142000	12731083	12.7504	8.5512	0.30944

Field	Q81	CUMQ	DELTAQ	T0	T1	PTF
Long Beach Airport	69225	1197000	48428	26.5336	24.8173	0.04046
Los Angeles City	51190	394000	14962	12.0056	11.2602	0.03798
Los Angeles Downtown	280812	3167000	153424	16.5456	15.3388	0.04844
Los Angeles, east	36632	302000	15996	11.8399	10.9215	0.05297
Los Lobos	13197	73861	1963	9.6192	9.1391	0.02657
Los Padres	366	223	15	0.8315	0.7559	0.06618
Lost Hills	5996238	97257000	4064896	24.6834	23.0524	0.04180
Lost Hills, northwest	3440	33219	4568	11.5606	9.6927	0.13751
Lyon Canyon	6345	105000	4345	25.2476	23.5902	0.04138
Mahala	50190	490000	24822	14.1698	13.1042	0.05066
McCool Ranch	682	1245	80	2.5127	2.2900	0.06387
McDonald Anticline	220102	838000	10268	8.2088	7.9351	0.01225
McKittrick	4719781	39117000	758331	15.5959	14.9428	0.01939
Midway-Sunset	43819980	582505000	72875903	16.1412	13.7299	0.12511
Monroe Swell	16935	109000	1770	12.7604	12.2728	0.01624
Montalvo, west	64099	483000	26119	10.7693	9.9218	0.05408
Montebello	486281	3254000	201606	9.2726	8.4688	0.06196
Moorpark, west	16442	102000	5735	8,7867	8.0761	0.05622
Morales Canyon	2442	19177	1269	10.7274	9.7522	0.06615
Mount Poso	9194702	83578000	2060929	15.9562	15.1934	0.02466
Mountain View	442599	5116000	191319	18.1023	16.9894	0.03740
Newgate	5346	35524	1471	10.1367	9.4710	0.04140
Newhall	34516	281000	20374	10.9080	9.8468	0.07250
Newhall-Potrero	348748	2417000	36252	14.0686	13.5519	0.01500
Newport, west	1177469	24210000	885658	32.3868	30.4233	0.03658
Oak Canyon	309308	1491000	875677	5.0657	2.0295	0.58731

Oak Park	51924	752000	33117	21.7517	20.2639	0.04404
Oakridge	158120	969000	417291	6.5486	3.6202	0.43064
Oat Mountain	103855	438000	26094	5.8954	5.3985	0.05957
Oil Creek	3100	8545	15	10.1197	9.9495	0.00180
Ojai	1335207	11031000	583976	11.8664	10.9463	0.05294
Olive	44847	298000	298000	6.8056	0.0000	1.00000
Orcutt	1185294	17724000	60978	46.4567	45.5095	0.00344
Oxnard	276711	2635000	227936	12.3202	10.9495	0.08650
Paloma	97195	1122000	222577	13.1807	10.2660	0.19838
Pioneer	1092	4849	70	9.1098	8.7810	0.01447
Piru	3157	19276	640	9.8691	9.3061	0.03320
Placerita	516295	3577000	196898	9.8610	9.0753	0.05505
Playa Del Rey	34147	448000	20549	19.5065	18.1354	0.04587
Pleasant Valley	6237	63971	11699	11.8228	9.3873	0.18289
Pleito	597004	6108000	305181	14.8984	13.7885	0.04996
Point Conception	10104	75180	4245	10.5281	9.6739	0.05647
Poso Creek	1812184	14183000	1742203	9.5301	8.1277	0.12284
Potrero	78232	376000	9967	8.2659	7.8539	0.02651
Prado-Corona	7448	24172	121	9.0841	8.8735	0.00502
Pyramid Hills	80985	696000	31639	12.8063	11.9115	0.04546
Quinado Canyon	407	5092	5092	12.7451	0.0000	1.00000
Railroad Gap	108745	1077000	54332	14.3887	13.3097	0.05045
Raisin City	212868	2384000	115760	16.4211	15.2215	0.04856
Ramona	151914	1684000	245803	13.1580	10.9229	0.14596
Richfield	1676772	16108000	319474	17.9572	17.1960	0.01983
Rincon	1410241	10421000	123225	16.1024	15.5743	0.01182
Rio Bravo	398754	1147000	71053	3.9860	3.6406	0.06195
Rio Viejo	543260	2254000	134549	5.7972	5.3079	0.05969
Riverdale	107496	954000	3654	26.7804	26.2149	0.00383

Field	Q81	CUMQ	DELTAQ	T0	T1	PTF
Rosecrans	298401	4000000	3371	58.5489	57.7336	0.00084
Rosecrans, east	4634	39158	3264	11.0106	9.8200	0.08336
Rosecrans, south	37429	344000	24875	12.3212	11.1249	0.07231
Rosedale	98266	801000	42542	11.6991	10.7900	0.05311
Rosedale Ranch	114249	800000	43941	9.9714	9.1781	0.05493
Round Mountain	432835	4299000	129915	16.4703	15.5831	0.03022
Russel Ranch	202574	2160000	188630	13.7705	12.2273	0.08733
Salt Lake	198984	3900000	146995	30.6316	28.7388	0.03769
Salt Lake, south	226632	2932000	135220	19.2094	17.8543	0.04612
San Ardo	10229425	171475000	7049860	25.6172	23.9427	0.04111
San Emidio Nose	78277	234000	176	13.3796	13.1981	0.00075
San Joaquin	12683	230000	9069	28.0121	26.2302	0.03943
San Miguelito	2246413	21731000	1103730	14.0315	12.9743	0.05079
San Vicenti	562892	9000000	378793	24.2891	22.6768	0.04209
Sansinena	428698	4675000	2378	52.8688	52.2091	0.00051
Santa Clara Avenue	288607	2275000	121781	11.2927	10.4103	0.05353
Santa Fe Springs	820986	15970000	1820498	23.9707	20.6520	0.11399
Santa Maria Valley	2997702	57329000	2192747	29.7747	27.9175	0.03825
Santa Paula	8953	101000	16757	13.1621	10.6692	0.16591
Santa Susana	105824	524000	4254	12.0642	11.7281	0.00812
Sargent	14972	157000	7786	15.2975	14.1638	0.04959
Saticoy	122136	904000	26833	12.3346	11.6773	0.02968
Saugus	7050	26695	240	8.9508	8.6905	0.00900
Sawtelle	282167	1680000	98517	8.3524	7.6564	0.05864
Seal Beach	994854	8113000	2469506	8.9323	6.0344	0.30439
Semitropic	189166	1275000	37462	11.2639	10.6673	0.02938

Sespe	1403946	6987000	1620142	5.5881	4.1698	0.23188
Seventh Standard	11641	50562	688	9.0761	8.7581	0.01361
Shiells Canyon	363151	5456000	1735681	16.3979	10.8589	0.31812
Simi	27675	173000	6923	9.6193	9.0015	0.04002
Sisquoc Ranch	4192	27876	1156	10.1391	9.4724	0.04148
South Mountain	1098123	7936000	943401	8.8456	7.5784	0.11888
Strand	80326	387000	22656	6.7613	6.1986	0.05854
Sunset Beach	13375	113000	113000	8.4997	0.0000	1.00000
Tapia	28976	380000	17433	19.4976	18.1270	0.04588
Tapo Canyon, south	50629	452000	2386	24.6387	24.0573	0.00528
Tapo Ridge	6235	45974	2530	10.4954	9.6593	0.05503
Tapo, north	14345	100000	5178	18.0654	9.2969	0.05178
Taylor Canyon	1746	4502	3	11.7178	11.5614	0.00070
Tejon	231548	2657000	53592	21.3446	20.4315	0.02017
Tejon Hills	55459	478000	45807	10.9420	9.6234	0.09583
Tejon, north	150737	1743000	83730	16.9988	15.7660	0.04804
Temblor Hills	255	501	32	2.7089	2.4695	0.06361
Temblor Ranch	6549	53434	3977	10.8734	9.7946	0.07443
Temescal	50163	679000	24894	21.3087	20.0150	0.03666
Ten section	142176	741000	42884	7.3333	6.7279	0.05787
Timber Canyon	79087	1195000	44828	23.6441	22.1875	0.03751
Torrance	1833269	15244000	2017808	10.0105	8.4436	0.13237
Torrey Canyon	184033	190000	94672	15.0439	13.9253	0.04983
Turk Anticline	12339	186000	8047	22.7409	21.2036	0.04326
Union Avenue	25181	208000	1276	21.8377	21.2929	0.00613
Union Station	37123	407000	19901	16.0482	14.8702	0.04890
Vallecitos	50314	584000	28018	17.0687	15.8320	0.04798

Field	Q81	CUMQ	DELTAQ	T0	T1	PTF
Valpredo	2741	27141	4175	11.6686	9.5964	0.15384
Venice Beach	36066	100000	58011	2.9155	1.1884	0.58011
Ventura	6971650	56934000	4436008	10.7831	9.6757	0.07791
Wayside Canyon	39647	292000	121573	7.8867	4.4695	0.416346
Welcome Valley	543	687	45	1.7360	1.5804	0.064923
West Mountain	37418	515000	11772	24.6904	23.5594	0.022857
Westhaven	19498	34184	2188	2.4121	2.1980	0.064007
Wheeler Ridge	413160	3865000	198154	13.5387	12.5120	0.051269
White Wolf	26289	597000	95909	26.6069	21.7044	0.160652
Whittier	566813	5600000	282708	14.3513	13.2745	0.050484
Wilmington	14256089	41052000	9023537	3.2513	2.4643	0.219807
Yorba Linda	2688001	23406000	2109571	11.1792	9.8952	0.090129
Yowlumne	4715868	19597000	1169588	5.8065	5.3166	0.059682
Zacca	627836	8973000	397428	21.4346	19.9629	0.044292
Belmont Offshore	1474448	31563000	1124506	33.9550	31.9306	0.035627
Elwood Offshore	93743	1160000	54400	18.2981	16.9926	0.046897
Elwood, South, offshore	2731937	10193000	636386	5.1614	4.7115	0.062434
Huntington Beach Offshore	7749338	83600000	623848	26.9431	26.2177	0.007462
Montalvo, west offshore	31011	100000	19224	3.6952	2.9002	0.192241
Newport, west offshore	8763	328000	6721	69.3050	66.3150	0.020490
Point Conception offshore	20782	156000	8443	10.7261	9.8815	0.054123
Rincon Offshore	324784	1558000	213111	5.7471	4.8226	0.136785
Summerland Offshore	120322	525000	24333	6.4708	6.0127	0.046348
Torrance Offshore	189786	1100000	11863	12.9816	12.5732	0.010785
Venice Beach Offshore	28279	718000	22522	41.6760	39.3798	0.031367
Wilmington Offshore	25209814	327388000	15076628	19.2895	17.9301	0.046051

	Q81	CUMQ	DELTAQ	T0	T1	PTF
Alegria Offshore	39316	201000	6641	8.2742	7.8036	0.033042
Carpenteria Offshore	1131993	12800000	619550	16.5924	15.3829	0.048402

Variable Definitions:

Q81 = production in 1981, bbl.

CUMQ = total production over economic lifetime of the field.

DELTAQ = Reduction in output due to the tax.

T0 = estimated lifetime of the field without the tax.

T1 = estimated lifetime of the field with the tax.

PTF = DELTAQ/CUMQ.

TABLE C.2
California Drilling Activity
(*t*-ratios in parentheses)

Independant Variables	Dependent Variable	
	Exploratory Holes	Completed Producers
Constant	3.3423	4.6988
P_t	0.4295	0.4164
	(3.2924)	(1.9751)
W_{t-1}	0.6730	0.6183
	(6.5988)	(4.7768)
t	−0.0133	−0.0232
	(−3.0806)	(−2.5534)
$(L/H)_t$	—	−0.2459
		(−2.5780)
WPT_t	−0.1158	−0.0827
	(−0.7392)	(−0.3480)
R^2	0.9141	0.5660
N	34	34

reflects imposition of the federal WPT. Thus the final equations estimated are of the form

$$W_{it} = b_{i0} + b_{i1}P_t + b_{i2}W_{i,t-1} + b_{i3}t$$

$$+ b_{i4}WPT_t + b_{i5}(L_t/H_t) + u_{it} \qquad (C.29)$$

where *i* stands for exploratory holes or completed producing wells, W_t is
the number of wells drilled in year *t*, L_t and H_t are the number of light and
heavy wells completed in year *t*, and WPT_t is a dummy variable equal to
1 during 1979 to 1982 and zero otherwise. As noted in chapter 5, b_5 was
set equal to zero in the exploratory holes regression. The variables W_t,
P_t, W_{t-1}, and (L_t/H_t) are in natural logarithms so that their coefficients
have an elasticity interpretation.

Empirical estimates of the drilling models are presented in table C.2. It
is evident from this table that all the coefficients have the expected sign,
and all but the WPT dummy are statistically significant. The WPT
variable was retained in the final specification despite its large standard
error because it belongs in the equations on theoretical grounds and
because the sign and magnitude of its estimated coefficients are quite
plausible.

Given these estimated models of drilling activity, it remains to project levels of drilling into the future with and without the tax. This was accomplished by inserting net price levels with and without the severance tax into the implied difference equations in W_{it} and then computing solution values over the forecast period. The 1981 real crude oil price was used for the baseline projection, reflecting an assumption of no real net price change, an appropriate assumption for the baseline projection. In after-tax projections price was reduced by 6 percent. (As explained in appendix *B*, the correct percentage reduction in net price is simply the nominal severance tax rate.)

Solving the difference equations is complicated by two factors: (1) Both equations contain a trend and (2) the WPT dummy variable runs only through 1990. In terms of our price expectation model [see equation (C.27)] this reflects an assumption that the WPT will be in effect through 1991 and will lapse in 1992. The difference equations are linear with constant coefficients, so the solution of each is given by the sum of the general solution to the associated homogeneous equation plus a particular solution (Goldberg 1958). Future values of L_t/H_t were forecast on the basis of its past trend from the estimated equation

$$ln(L_t/H_t) = 4.1623 - .0739t \quad N = 35 \quad R^2 = .7968$$

$$(9.8870)(-11.3749) \quad (t\text{-}ratios \; in$$

$$parentheses) \quad (C.30)$$

This trend was incorporated into the constant and trend terms in the equation for completed producing wells for purposes of forecasting drilling activity.

Thus the equation for W_{it} is a difference equation of the form (omitting the *i* subscript)

$$W_t = a_0 + a_1P + a_2W_{t-1} + a_3t + a_4D_t \quad (C.31)$$

where P equals the real net price either with or without the tax and $D_t = 1$ through 1990, zero afterwards. The general solution to the homogeneous equation is

$$W_t = Ca_2^t \quad (C.32)$$

where C is a constant to be determined by initial conditions. Since a_2 is positive and less than one, the contribution of this part of the solution fades out in the long run. The method of undetermined coefficients yields a particular solution, so that the full solution for W_t is

$$W_t = [a_0/(1-a_2)] + [a_1 P/(1-a_2)] + [a_3 t/(1-a_2)]$$
$$- [a_2 a_3/(1-a_2)^2] + [a_4/(1-a_2)]a_2^{(1-D_t)(t-t*)}$$
$$+ Ca_2^t \tag{C.33}$$

where $t* = 90$, the last (two-digit) year in which the WPT variable takes a nonzero value. The constant C is determined by the initial condition that $W_{82} =$ the fitted value of the estimated equation for $t = 82$.

Forecast values of W_t for exploratory holes and completed producers over a 30-year period after imposition of the tax are reported in table C.3. It is of interest to note that the long-run elasticity of drilling with respect to net price is given by $a_1/(1 - a_2)$. The value of this elasticity for exploratory drilling is 1.313 and for development drilling it is 1.091. When compared to the independent crude oil supply elasticity estimates reported in table 4.4. in the text, it is quite clear that our estimates are consistent with the findings of other authors.

Drilling projections can be translated into projections of lost output by estimating the initial production from new wells in California over the 30-year forecast period. Using data from the Conservation Committee of California Oil Producers (1982) for the period 1957 to 1981, the trend in initial daily output per new well was estimated as

$$\ln Q_{it} = 6.6079 - .0360t \quad N = 25 \quad R^2 = .5056$$
$$(12.8833)(-4.8503) \quad (t\text{-}ratios\ in$$
$$parentheses) \tag{C.34}$$

The trend is downward, reflecting the fact that the most productive wells tend to be drilled first. Initial output from the new producing wells drilled over the 30-year projection period was extrapolated from equation (C.34). This quantity, multiplied by the forecast number of new producing wells drilled (and the result multiplied by 365 to transform the result into an annual figure), constitutes our estimate of the initial production for wells drilled in a given year.

To estimate total production from new wells in each year required information on (1) the decline rate of the new wells and (2) the expected lifetime of the new wells without the severance tax. Here we assumed new wells would exhibit the same decline rate presently observed in the state as a whole. However, since our estimate of the state average decline

TABLE C.3
Exploratory Holes Drilled per Year (EXHNEW), Completed Producing Wells
Drilled per Year (CWSNEW), Total New Wells Drilled per Year (WNEW), 6%
Severance Tax, First 30 Years after Imposition of Tax

Year	EXHNEW[a]	EXHNEWT[b]	CWSNEW[c]	CWSNEWT[d]	WNEW[e]	WNEWT[f]
1	541	518	3119	2992	3520	3375
2	551	520	3312	3146	3720	3531
3	549	515	3420	3229	3827	3610
4	542	505	3471	3264	3872	3638
5	529	492	3485	3270	3877	3634
6	514	477	3477	3257	3857	3610
7	498	461	3454	3233	3822	3574
8	481	444	3423	3202	3778	3531
9	520	480	3678	3440	4063	3795
10	541	500	3826	3578	4227	3948
11	549	506	3901	3648	4307	4022
12	546	504	3929	3673	4333	4046
13	538	496	3926	3670	4324	4037
14	525	484	3904	3650	4293	4008
15	510	470	3871	3619	4249	3967
16	493	455	3832	3582	4197	3918
17	476	439	3789	3541	4141	3866
18	459	423	3743	3499	4083	3812
19	442	407	3697	3455	4023	3757
20	425	392	3650	3411	3964	3701
21	408	377	3603	3368	3905	3646
22	392	362	3556	3324	3847	3592
23	377	348	3510	3281	3789	3538
24	362	334	3464	3238	3732	3485
25	348	321	3419	3196	3676	3433
26	334	308	3374	3154	3621	3382
27	321	296	3330	3112	3567	3331
28	308	284	3286	3072	3514	3282
29	296	273	3243	3031	3462	3233
30	284	262	3201	2992	3411	3186

[a] Without tax.
[b] With tax.
[c] Without tax.
[d] With tax.
[e] Without tax.
[f] With tax.

TABLE C.4
Annual Production from New and Existing Wells, Millions of Barrels, 6 Percent
Severance Tax, First 30 Years after Imposition of Tax

Year	ANNQ[a]	ANNQT[b]
1	293.9	292.2
2	317.6	314.0
3	338.8	333.1
4	356.9	349.0
5	371.1	359.7
6	380.3	367.6
7	387.1	374.2
8	394.0	380.1
9	401.1	375.7
10	405.7	377.5
11	408.3	378.1
12	406.4	379.9
13	401.8	381.8
14	399.8	372.3
15	397.9	348.3
16	394.3	343.0
17	362.7	336.2
18	352.3	330.8
19	344.6	326.7
20	339.8	322.3
21	334.4	314.2
22	327.0	308.5
23	319.8	296.3
24	313.3	275.8
25	294.6	239.3
26	283.9	233.4
27	274.5	226.7
28	266.8	220.2
29	259.8	212.8
30	246.6	206.5

[a] Without tax.

[b] With tax.

rate varied depending on whether the average was weighted (by reserves) or unweighted, we generated two sets of output estimates. Regarding new well lifetimes we assumed that within the 30-year forecast period, none of the newly drilled wells would be prematurely shut in as a result of the tax. In terms of the size of estimated production effects, this assumption is of course quite conservative. Final estimates of the total production impact of the tax are presented in tables C.4 and C.5. Table C.4 gives total annual production with and without the tax; cumulative production losses from the tax and the production loss expressed as a percent of

projected cumulative output appear in table C.5. These data are plotted
in chapter 5.

Two additional features of the data used to estimate production effects
deserve mention. Since solutions of the difference equations used to
forecast drilling are explicitly dynamic in form, we were able to begin our

TABLE C.5
Cumulative Production Losses

Year	DELTAQT[a]	DELQTOT[b]	PTAX[c]
1	0.243	1.746	0.59
2	0.534	5.302	0.87
3	1.139	10.987	1.16
4	2.210	18.877	1.44
5	5.236	30.319	1.81
6	8.090	43.007	2.09
7	9.874	55.853	2.28
8	11.659	69.746	2.46
9	23.870	95.099	2.93
10	37.979	123.305	3.38
11	53.239	153.500	3.79
12	64.032	179.935	4.03
13	67.899	200.015	4.11
14	78.763	227.533	4.32
15	111.379	277.124	4.90
16	145.506	328.437	5.42
17	154.706	354.936	5.53
18	158.931	376.486	5.56
19	159.561	394.390	5.54
20	159.855	411.839	5.52
21	163.102	432.067	5.55
22	164.853	450.574	5.55
23	171.870	474.080	5.62
24	193.142	511.538	5.85
25	232.630	566.881	6.27
26	267.619	617.368	6.62
27	300.380	665.252	6.93
28	332.246	711.819	7.21
29	364.829	758.761	7.49
30	391.065	798.914	7.70

[a] Cumulative loss in production from existing wells in millions of barrels.

[b] Cumulative loss in production from all wells in millions of barrels.

[c] Percentage loss in output from all wells due to 6-percent severance tax, first 30 years after imposition of tax.

simulations of drilling effects as of 1983. Thus year 1 in table C.3 and elsewhere corresponds to 1983. The most recent reserves data available, however, are from DOG *Annual Report* 1981. We began counting the existing well effect as of the date of the last available reserves estimate, 31 December 1981. This amounted to assuming that existing fields at the end of 1982 were no different than they were at the end of 1981. Since existing wells were in fact one year older at the end of 1982, our procedure will slightly understate the effect of the tax on old wells and fields.

In estimating effects of small producer exemptions, our source of data lists joint and unitary operations separately from individual production operations. For purposes of estimating the taxable production per field, joint or unitary operations involving a particular company were attributed to a different company than production from the same firm's individual operations. This procedure has the effect of slightly understating the amount of taxable production for some companies that have scattered joint, unitary, and individual operations. Accordingly, it will tend to overstate the effect of the exclusions.

Chapter 6

Impacts of a Crude Oil Severance Tax on Employment, Sales, and Government Revenues

Imposing a 6-percent severance tax on crude oil produced in California can be expected to curtail business activity in the crude oil production and drilling industries and in the industries that supply inputs to them. As a consequence demands for labor, capital, and other factors of production will decline and incomes of these factors will be lowered. These effects would be most heavily concentrated among inputs used directly in oil production and drilling. In addition profits of firms engaged directly in oil production and drilling will be reduced, as will profits of firms that supply materials and other factors of production to the oil production and drilling industries. Moreover, all such effects will be concentrated geographically in those regions where direct production and drilling effects are largest.

These general observations indicate how the general burden, or incidence, of the proposed severance tax is expected to be distributed. A new tax, or increase in a tax rate, typically will impose burdens both on consumers of the taxed item and on those who produce it or supply inputs to its production. The burden on consumers comes in the form of higher after-tax prices, while burdens on the supply side of the market are reflected in lower after-tax profits and factor incomes. The proposed severance tax is not expected to have any effect on crude oil prices, however. Thus the entire burden of the tax is expected to be on the supply side of the market, among firms and suppliers of inputs to the crude oil production and drilling industries. This conclusion follows from the view that crude oil prices in California and elsewhere are determined in a worldwide market. Since California produces only a tiny fraction of worldwide supply, it is anticipated that the projected output reductions in California will have no discernible effect on crude oil prices.

The remainder of this chapter analyzes these supply side effects. Our aim is not to provide a complete incidence analysis, that is, estimates of reductions in profits and factor incomes throughout the state. Such an undertaking is outside the scope of the present study. Rather our intent is to indicate only the general pattern of incidence by estimating tax-induced reductions in output and employment demand for various industries in the California economy.

It should also be noted that estimated reductions in output and employment demand presented are attributable solely to tax-induced reductions in crude oil production and drilling. No attempt has been made to assess the output and employment effects of the resulting increase in state tax revenue. This revenue would presumably be used either to increase state government spending or reduce collections from other state taxes. Questions relating to "balanced budget" or "differential" incidence are, however, beyond the scope of our analysis.

The impact estimates reported are derived from estimated reductions in drilling and production presented in chapter 5. As analysis in that chapter demonstrated, exempting small oil producers (less than 36,500 bbl./yr. or 100,000 bbl./yr.) will not alter production and drilling effects of the tax to any substantial degree unless these induce a significant change in the structure of the industry. Because we lack a firm basis to judge the likelihood or magnitude of any such structural change, estimates presented in this chapter assume that no exemptions are granted.

6.1 Market Effects of a Crude Oil Severance Tax

Our view of the market effects of a crude oil severance tax in California parallels conclusions reached in a recent analysis of energy taxation issues by the U.S. Department of Energy (DOE): "In sum, the effects of a state severance tax on oil will be to lower domestic oil output, to increase oil imports, and to lower incomes of factors employed in the oil industry, but it will not have a significant effect on crude oil prices" (U.S. Energy Information Administration 1980, p. 51). This follows from the importance of both interstate and international trade in crude oil. Any attempt by local producers to include state production taxes in prices they charge refiners will render their oil noncompetitive in relation to imports. Because the supply of imported crude oil to the California market is highly elastic, imports could be increased to offset local production declines without raising the prices paid by refiners.

This line of reasoning is summarized in figure 6.1. The curve labeled D is demand for crude oil by California refiners, and S_0 represents the pretax supply schedule of local producers. Assuming that imports are

FIGURE 6.1
California Crude Oil Market with
and without a Severance Tax

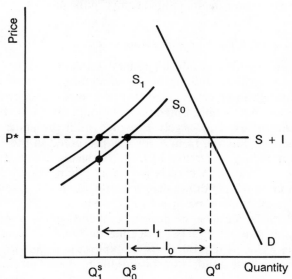

perfectly elastic at price P^*, the total supply of crude oil to refiners in the state is seen to be $S+I$. Hence in the pretax situation the equilibrium price is P^*, domestic production is Q_0^s, refiners demand Q^d barrels, and imports are I_0. When the tax is imposed domestic supply is reduced to S_1. As a consequence domestic production falls to Q_1^s, and imports rise to I_1. Note that the fall in domestic production is just offset by the increase in imports. The total amount supplied to refiners and the equilibrium price both remain unchanged.

From this view of the crude oil market it follows that the entire burden of the tax falls on the supply side of the market. As pointed out in the DOE study we cited, "State and local production taxes could be expected to be shifted backward to the quasi rents that accrue to owners of capital in crude oil production, to labor, and to owners of mineral rights" (U.S. Energy Information Administration 1980, p. 50). Among these factors of production, the burden of the tax would be felt either as reduced levels of employment, reduced factor payments (prices and wages), or perhaps both. An important part of the analysis in this chapter involves estimating the effects of a severance tax on employment demand in the crude petroleum industry in California.

6.2 Macroeconomic Effects of a Severance Tax

At present about half the crude oil refined in California is imported, primarily from Alaska and Indonesia. As pointed out, a crude oil severance tax would tend to divert demand from domestic production and to imported supplies. Accordingly, income that would otherwise have been earned by Californians, and largely respent within the state, will now accrue to outsiders. Judging from estimates presented in the preceding chapter, the tax-induced increase in imports during the first 10 years would average 12.33 million barrels per year. When valued at current prices, the amount of expenditure diverted from domestic producers and to imports is $330 million per year. (See table 6.2.)

Any such disturbance to a single sector of the state's economy will have a variety of indirect effects. A decline in domestic production of crude oil, plus associated reductions in exploration and development activity, will reduce demands that oil-producing firms place on other industries in the state. Thus those who supply steel tubing, pumping equipment, transportation services, and so forth, to the crude oil sector will experience a general decline in business activity. Similarly, reductions in wages earned by crude oil workers imply fewer expenditures for household consumption. These first-round effects then generate successive impacts that ripple through the economy.

The nature of these secondary effects is perhaps best understood in terms of an example. In table 6.1 the first column of figures reflects the pattern of purchases made by the oil- and gas drilling industries in California. Individual entries represent amounts purchased from various sectors of the state economy, per million dollars of total expenditure on drilling oil and gas wells. Thus for each million dollars spent drilling new wells, the drilling industry buys $3,000 of output from agriculture, forestry, fisheries; $8,000 from mining; $205,000 from manufacturing; and so forth. The entry attributed to households represents $415,000 in payments to labor. Overall the original $1 million generates $785,000 in first-round purchases from other industries in the state. These initial indirect purchases then give rise to second-, third-, and so forth, round changes in economic activity. The total of all these indirect and induced changes in total sales, by sector, is shown in the second column of table 6.1. Thus the initial $1 million change in drilling expenditure generates

TABLE 6.1

Illustrative Impact of $1 Million Increase in Drilling Activity on Gross Sales, by Industry (millions of 1976 dollars)

	First-Round Sales	Total Indirect and Induced Sales
Agriculture, forestry, fisheries	0.003	0.040
Mining	0.008	0.039
Construction	— *a*	0.023
Manufacturing	0.205	0.643
Wholesale trade	0.021	0.087
Retail trade	0.028	0.200
Transportation, communication, utilities	0.023	0.175
Finance, insurance, real estate	0.010	0.278
Services	0.063	0.300
Households	0.415	0.997
Other	0.008	0.070
Total	0.785	2.852

SOURCE: California Department of Water Resources (1980).
NOTE: All impacts valued at 1976 prices.
a Less than 0.0005.

$643,000 in increased manufacturing production, $200,000 in increased retail trade, $997,000 in increased payments to labor, and so forth. Overall these indirect and induced impacts sum to $2.852 million; if added to the $1 million *direct* effect, the *total* impact of this $1 million change in drilling expenditure is $3.852 million. Thus the total impact on sales is a multiple (3.852) of the direct effect.

6.3 Application of the Input-Output Approach

The technique used to obtain these results is commonly known as input-output analysis. This paradigm provides a simple, empirically tractable framework for studying equilibrium relationships among sectors in an economy. (For a more detailed explanation of input-output theory, and its application to the present problem, see appendix *D*.) The estimates presented in table 6.1 and the more detailed results that follow were obtained from an input-output model constructed by the California Department of Water Resources (DWR) in 1980. The DWR model was explicitly constructed to facilitate the kind of economic impact analysis undertaken here. Because the model was based on 1976 data, it was necessary to modify and update it to reflect changes in prices and labor productivity. These modifications are fully explained in appendix *D*.

The input-output approach has been widely used to estimate economic impacts (see Richardson 1972 for applications). It is appropriate to point out some of its limitations, however, to provide a proper context for interpreting estimates in this chapter. Input-output is designed to provide detailed estimates of the amounts of labor and other inputs required to produce a given set of outputs in an economy. Accordingly, the employment impacts presented in this chapter are best termed changes in labor requirements or employment demands. The extent to which these shifts result in actual unemployment will depend on several factors, including the future course of general unemployment and macroeconomic activity in the state, the ability of employers to substitute between labor and other inputs, and the degree to which labor supply readjusts through relocation, retraining, and so forth. None of these factors can be forecast beyond a relatively short time horizon.

Furthermore, as already noted, the following analysis examines only those impacts attributable to reductions in crude oil production and drilling activity (and the associated output reductions in other sectors caused by multiplier effects). Beyond these production effects, revenue raised by the severance tax would represent a transfer of income from the oil industry to the government. The impact estimates that follow ignore effects that this pure transfer might have on aggregate economic activity in the state. Any attempt to estimate macroeconomic impacts from this transfer would require forecasts of how government would spend the proceeds of the tax, plus estimates of the effect of reduced industry income on spending patterns and final demands in the state. It is clear that part of the industry's tax payment would be offset by reduced income tax liabilities, and this would have little or no effect on final demand. However, the magnitude of this offset and the degree to which the remainder of the transfer from industry would alter final demand are unknown. On the government's side of the transfer, it would be necessary to assume an expenditure pattern for the severance tax revenue and then analyze the economic stimulation associated with it. All of these issues are quite separate from the effect of the tax on crude oil production in the state, and analysis of them is outside the scope of the present study. Accordingly, the pure transfer aspect of the tax is not considered in the estimation of output and employment impacts; the only changes in

TABLE 6.2
Direct Effects of Severance Tax

	10-Year Average	20-Year Average	30-Year Average
Drilling effect: decrease in			
total wells drilled (per year)	232	253	250
Drilling expenditure	148.048	162.713	159.534
Production effect (9-percent decline rate) decrease in			
Crude oil production (mmbbl. per year)	12.330	20.592	26.630
Value of sales ($ millions per year)	329.458	550.218	711.554
Production effect (3-percent decline rate) decrease in			
Crude oil production (mmbbl. per year)	13.758	25.147	34.313
Value of sales ($ millions per year)	376.614	671.928	916.843

SOURCE: See appendix *D* and chapter 3.
NOTE: Dollar amounts are in millions, at 1982 prices.

employment demand estimated are those arising from the effect of the severance tax on production.

To implement the input-output approach it is necessary to represent the direct impact of the severance tax as a change in final demand in the economy. Within the DWR framework, tax-induced production effects would occur in two sectors, "crude petroleum," which represents the operation of oil properties, and "new construction, all other," which incorporates the drilling of oil and gas wells. The estimated effect of the severance tax on levels of outputs in these sectors is shown in table 6.2. Recall from chapter 5 that two sets of production effects (associated with new oil well production decline rates of 3 and 9 percent per year) were computed. Accordingly, a range of final demand changes and resulting economic impacts is analyzed.

6.3.1 Effects on Gross Sales

The estimated effects of the tax on gross sales, valued at 1982 price levels, are shown in tables 6.3 and 6.4. Notice that the overall multiplier effect declines slightly over time as the importance of the production effect grows. In total the anticipated reduction in annual gross sales is estimated at over $2.0 billion for the full 30-year period. The industrial distribution of gross sales effects was also computed but is not reported here. These detailed industry-specific figures were used, however, in the estimation of impacts on employment and tax receipts.

TABLE 6.3

Effect of Tax on Annual Gross Sales: 9-Percent New-Well Decline Rate
(millions of 1982 dollars)

	10-Year Average	20-Year Average	30-Year Average
Production effect			
Direct	329.470	550.217	711.566
Indirect	333.582	557.083	720.444
Total	663.052	1107.300	1432.010
Gross sales multiplier	2.012	2.012	2.012
Drilling effect			
Direct	147.894	162.646	159.832
Indirect	424.400	466.731	458.658
Total	572.294	629.377	618.490
Gross sales multiplier	3.870	3.870	3.870
Total effect			
Direct	477.364	712.863	871.398
Indirect	757.982	1023.814	1179.102
Total	1235.346	1736.677	2050.500
Gross sales multiplier	2.588	2.436	2.353

SOURCE: See text.

TABLE 6.4

Effect of Tax on Annual Gross Sales: 3-Percent New-Well Decline Rate
(millions of 1982 dollars)

	10-Year Average	20-Year Average	30-Year Average
Production effect			
Direct	367.626	671.947	916.856
Indirect	372.214	680.333	928.294
Total	739.840	1352.280	1845.150
Gross sales multiplier	2.012	2.012	2.012
Drilling effect			
Direct	147.894	162.646	159.832
Indirect	424.400	466.731	458.658
Total	572.294	629.377	618.490
Gross sales multiplier	3.870	3.870	3.870
Total effect			
Direct	515.520	834.593	1076.688
Indirect	796.614	1147.064	1386.952
Total	1312.134	1981.657	2463.640
Gross sales multiplier	2.545	2.374	2.288

SOURCE: See text.

6.3.2 Effects on Employment

The estimated changes in gross sales by industry are used to forecast the employment effects of a severance tax. (See California Office of Planning and Research 1978, p. 50ff., for further discussion of this procedure.) The technique proceeds from the assumption that the ratio of employment to gross sales in each industry will remain constant when production and drilling effects of the tax occur. This is appropriate if the reduction in labor demand caused by the tax does not result in a change in the relative price of labor to other goods. If, on the other hand, wages were to fall as a result of tax-induced reductions in demand, then the ratio of employment to gross sales would be expected to increase somewhat. The question of wage flexibility is discussed later in the context of the occupational composition of these employment impacts.

Summary information on the effects of a severance tax on employment demand is presented in tables 6.5 and 6.6 for new well production decline rates of 9 and 3 percent, respectively. All employment demand effects are expressed in person-years of full-time equivalent employment. In each table, three columns of figures corresponding to different time spans are presented. (Note that these time spans are cumulative and hence overlap; for example, the 10-year average effects are included in estimates for both 20- and 30-year averages.) From these estimates it is apparent that the employment effect of reduced drilling is relatively constant and averages 1,200–1,300 direct and 5,400–6,000 total person-years of employment per year. Given the estimated drilling model presented in chapter 5, this constancy was not unexpected. From the nature of the lagged response estimated for the drilling industry, reactions to net price changes take place quite quickly; only about three years are required for the industry's response to the new tax to be 80 percent complete. Notice also that the effect of reduced drilling on total employment is about 4.5 times as large as the direct effect, a multiplier that exceeds the drilling sector's multiplier for gross sales. This occurs because the drilling industry is somewhat less labor intensive than the industries that supply it.

In contrast to drilling impacts, employment effects of reduced production activities tend to build up over time. The direct impact of reduced production on employment demand rises from an average of about 500 direct jobs over the first 10 years to an average of about 1,000 direct jobs over the first 30 years. The total impact rises from an average of approximately 3,600 jobs over the first 10 years to an average of almost 8,000 jobs over the first 30 years. (These impacts are for a 9-percent decline rate for new wells; larger employment impacts result from a 3-percent decline rate as table 6.6 shows.) Recalling that the 20- and 30-year average figures are cumulative, the actual acceleration in employment impacts is more dramatic than it might appear from tables 6.5 and 6.6. In the case of production responses, indirect employment effects are far greater than direct impacts. For each direct job loss to the crude petroleum production sector, about 6.7 indirect employment opportunities disappear. This large multiplier effect results from the fact that the crude oil production industry is far less labor intensive than the industries that sell to it. Of the producing sectors included in the DWR input-output model, only one is less labor intensive than crude oil production.

Though not reported separately, employment effects for shorter time spans were also computed. As already noted, the drilling response to the tax is expected to be quite rapid. Direct and indirect employment reductions attributable to reduced drilling are expected to average about 4,300 jobs during the first three years and 4,900 over the first five years (assuming a 9-percent new-well decline rate). From table 6.5 it can be seen that this is not substantially different than the estimated 10-year average. Production employment impacts, on the other hand, grow more slowly. This is due to the fact that the impact of the tax on production from existing wells and fields tends to increase as the fields age, as discussed in chapter 5. Over the first three- and five-year periods, total employment demand reductions attributable to reduced production are

TABLE 6.5

Tax-Induced Reductions in Employment Demand: 9-Percent New-Well Decline Rate
(full-time equivalent jobs)

	10-Year Average	20-Year Average	30-Year Average
Production effect			
Direct	471	787	1,018
Indirect	3,171	5,295	6,848
Total	3,642	6,082	7,866
Drilling effect			
Direct	1,208	1,329	1,306
Indirect	4,248	4,672	4,591
Total	5,456	6,001	5,897
Total effect			
Direct	1,679	2,116	2,324
Indirect	7,419	9,967	11,439
Total	9,098	12,083	13,763

SOURCE: See text.

TABLE 6.6

Tax-Induced Reductions in Employment Demand: 3-Percent New-Well Decline
Rate
(full-time equivalent jobs)

	10-Year Average	20-Year Average	30-Year Average
Production effect			
Direct	526	961	1,311
Indirect	3,538	6,467	8,824
Total	4,064	7,428	10,135
Drilling effect			
Direct	1,208	1,329	1,306
Indirect	4,248	4,672	4,591
Total	5,456	6,001	5,897
Total effect			
Direct	1,734	2,290	2,617
Indirect	7,786	11,139	13,415
Total	9,520	13,429	16,032

SOURCE: See text.

1,100 and 1,800, respectively, or about 30–50 percent of the 10-year average.

Tables 6.7–6.12 present the industrial composition of anticipated employment impacts. In interpreting these figures, note that crude oil production employment is included in the mining sector, while drilling employment is incorporated in construction. Estimates for the two different production decline rates are sufficiently similar that separate discussions are not warranted. During the first 10 years employment impacts are dominated by reduction in drilling activity. Accordingly, direct employment effects are concentrated in the construction sector. Additional reductions in overall employment are expected to be concentrated in retail trade, services, and manufacturing; of these, impacts in the first two sectors are traceable primarily to reductions in wage payments and household income in the state. In general the distribution of indirect employment effects is roughly similar to the industrial distribution of employment in the state as a whole. Over a longer time frame, employment effects of reduced production grow, and their industrial composition changes somewhat. According to our estimates the 30-year average employment effects are still concentrated in retail trade, services, and manufacturing. However, the share of direct job losses that

TABLE 6.7
Industry Composition of 10-Year Average Reduction in Employment Demand:
9-Percent New-Well Decline Rate
(full-time equivalent jobs)

Sector	Production Effect	Drilling Effect	Total Effect
Agriculture, forestry, fisheries	75	130	205
Mining	502	30	532
Construction	137	1,243	1,380
Manufacturing	450	753	1,203
Wholesale trade	199	333	532
Retail trade	760	1,173	1,933
Transport., communications, utilities	193	291	484
Finance, insurance, real estate	459	345	804
Services	797	1,081	1,878
Other	70	77	147
Total	3,642	5,456	9,098

TABLE 6.8
Industry Composition of 20-Year Average Reduction in Employment Demand:
9-Percent New-Well Decline Rate
(full-time equivalent jobs)

Sector	Production Effect	Drilling Effect	Total Effect
Agriculture, forestry, fisheries	124	143	267
Mining	838	33	871
Construction	229	1,367	1,596
Manufacturing	752	828	1,580
Wholesale trade	332	320	652
Retail trade	1,270	366	1,636
Transport., communications, utilities	322	1,290	1,612
Finance, insurance, real estate	767	380	1,147
Services	1,331	1,189	2,520
Other	117	85	202
Total	6,082	6,001	12,083

occur in mining has increased, while the portion attributable to construction has fallen.

To provide a check on the overall magnitudes of our estimated employment effects, direct losses in employment demand resulting from both drilling and production impacts were compared to independent data on total employment in all phases of oil and gas production in California. The results of this comparison are shown in table 6.13. Depending on the total employment figure used and the time span considered, *direct* job losses amount to 4.4–6.7 percent of total employment in this industry. For comparison the estimated percentage reductions in crude oil produc-

TABLE 6.9

Industry Composition of 30-Year Average Reduction in Employment Demand: 9-Percent New-Well Decline Rate
(full-time equivalent jobs)

Sector	Production Effect	Drilling Effect	Total Effect
Agriculture, forestry, fisheries	161	141	302
Mining	1,084	32	1,116
Construction	296	1,343	1,639
Manufacturing	973	813	1,786
Wholesale trade	429	360	789
Retail trade	1,642	1,268	2,910
Transport., communications, utilities	416	315	731
Finance, insurance, real estate	992	373	1,365
Services	1,721	1,169	2,890
Other	152	83	235
Total	7,866	5,897	13,763

TABLE 6.10

Industry Composition of 10-Year Average Reduction in Employment Demand: 3-Percent New-Well Decline Rate
(full-time equivalent jobs)

Sector	Production Effect	Drilling Effect	Total Effect
Agriculture, forestry, fisheries	83	130	213
Mining	560	30	590
Construction	153	1,243	1,396
Manufacturing	502	753	1,255
Wholesale trade	222	333	555
Retail trade	848	1,173	2,021
Transport., communications, utilities	215	291	506
Finance, insurance, real estate	513	345	858
Services	889	1,081	1,970
Other	79	77	156
Total	4,064	5,456	9,520

TABLE 6.11
Industry Composition of 20-Year Average Reduction in Employment Demand:
3-Percent New-Well Decline Rate
(full-time equivalent jobs)

Sector	Production Effect	Drilling Effect	Total Effect
Agriculture, forestry, fisheries	152	143	295
Mining	1,023	33	1,056
Construction	280	1367	1,647
Manufacturing	918	828	1,746
Wholesale trade	405	320	725
Retail trade	1,551	366	1,917
Transport., communications, utilities	393	1,290	1,683
Finance, insurance, real estate	937	380	1,317
Services	1,625	1,189	2,814
Other	144	85	229
Total	7,428	6,001	13,429

TABLE 6.12
Industry Composition of 30-Year Average Reduction in Employment Demand:
3-Percent New-Well Decline Rate
(full-time equivalent jobs)

Sector	Production Effect	Drilling Effect	Total Effect
Agriculture, forestry, fisheries	207	141	348
Mining	1,396	32	1,428
Construction	382	1,343	1,725
Manufacturing	1,253	813	2,066
Wholesale trade	552	360	912
Retail trade	2,116	1,268	3,384
Transport., communications, utilities	537	315	852
Finance, insurance, real estate	1,279	373	1,652
Services	2,217	1,169	3,386
Other	196	83	279
Total	10,135	5,897	16,032

tion are also shown in table 6.13; a close correspondence between percentage employment and production effects is evident. On the basis of this comparison, then, our estimates of reductions in direct employment appear quite reasonable.

Information on the occupational distribution of employment in "mining," which incorporates both extractive operations and drilling activity, together with the current occupational distribution of statewide employment, is presented in table 6.14. Recall that the industrial distribution of *indirect* jobs lost as a result of the tax roughly parallels the overall employment pattern for the state. Thus the two sets of figures in table 6.14 are indicative of occupations where direct and indirect employment impacts are expected to occur. (Estimating a complete occupational distribution of employment impacts was outside the scope of the present study.) Note first the heavy concentration of craftsmen and operatives in the mining industry. These occupations tend to be heavily unionized; hence wages are less flexible than would otherwise be the case. Consequently unemployment rather than downward wage adjustment is expected to be the primary result of a reduced demand for labor in these occupations. It is also noteworthy that labor market conditions in these occupations are already characterized by excess supply. Among the multitude of individual job categories in the broad classes of craftsmen and operatives now employed in mining, only maintenance mechanics were considered to be in high demand by the California Employment Development Department (1981) in a recent survey. Moreover, primary alternative employment opportunities for workers in these occupations are in construction and manufacturing industries (California Employment Development Department 1981), both of which are presently suffering substantial unemployment.

Thus the occupational distribution of unemployment impacts does not mesh well with current labor market conditions. However, we cannot say with any precision how long the present pattern of unemployment in California will persist.

In addition to industrial and occupational breakdowns, the regional

TABLE 6.13
Losses of Direct Jobs as a Percentage of Overall Extractive Employment in California
(9-percent production decline rate)

	10-Year Average (%)	20-Year Average (%)	30-Year Average (%)
Direct job losses as percentage of			
IPAA employment in 1981	4.4	5.6	6.1
EDD employment in 1981	4.9	6.1	6.7
Percent reduction in output	3.4	5.5	7.7

TABLE 6.14
Occupational Distribution of Employment, 1980
(percent)

	Mining (%)	All Industry (%)
Professional, technical, kindred	18.6	18.4
Managers, officials, proprietors	8.3	10.8
Sales workers	0.6	8.2
Clerical workers	11.7	19.5
Crafts and kindred workers	23.4	12.2
Operatives	34.5	12.1
Service workers	1.0	12.6
Laborers, except farm	2.0	3.5
Farmers and farm workers	N.A.	2.6
Total, all occupations	100.0	100.0

SOURCE: California Employment Development Department, *Annual Planning Information 1981–1982* (Sacramento, Calif., 1981).

distribution of employment impacts is also of interest. Though it is technically quite feasible, the time and resources available for the present study did not permit us to construct disaggregated regional estimates of the effect of the proposed tax on production and employment. It is, however, obvious that *direct* employment impacts would be most heavily felt in the major producing regions. At present about two-thirds of statewide employment in fossil fuel extraction is concentrated in Kern and Los Angeles counties (California Employment Development Department 1981). Most of the remainder is distributed among Ventura, Santa Barbara, and Orange counties.

When compared to the size of the current labor force in each county, it is clear that employment effects of the tax would be most severe in Kern County. Though we have not prepared precise regional estimates, it is possible using existing data to gain a rough impression of this regional impact. From information already presented, the proposed severance tax is expected to reduce direct employment demand by 4.5–6.7 percent. Suppose the employment effect in Kern County is typical of the statewide impact and a midpoint estimate of 5.6 percent is used. Within Kern County 8.5 percent of the labor force is directly employed in fossil fuel mining (California Employment Development Department 1981). Thus reductions in direct job opportunities would amount to about 0.48 percent of the local work force. According to estimates presented earlier, each direct job lost is accompanied by 4.4–4.9 indirect job losses. There is, of course, no reason to believe that all indirect job losses would be concentrated in the major oil-producing regions. According to analysis presented by the California Department of Water Resources (1980), however, it is not at all unreasonable to expect that at least 40 percent of the indirect effect would remain within the local region. (This implies that local indirect job losses would be about 1.76–1.96 times as large as direct losses.) Considering both direct and indirect effects then, the total loss in job opportunities in the county would amount to 1.3–1.4 percent of the local work force.

6.3.3 Effects on General Tax Revenues

One consequence of the decline in business activity expected to accompany a crude oil severance tax is a reduction in state and local government receipts from other taxes. For example, in table 6.1 it was shown that a $1 million decline in drilling activity reduces household incomes by about $1 million as well, and lowers retail trade volumes by $200,000 per year. Because the state government derives a major portion of its revenue from taxes on individual incomes and retail sales, it follows that any assessment of the net revenue attributable to the severance tax must take these secondary effects into account. These computations are the subject of the present section.

It is important to recognize that the tax interactions examined here are distinct from the interrelationships that arise from the simple fact that some taxes may be deducted from the bases of other taxes. The latter point was discussed in chapter 4 and appendix *B* (see also Camm et al. 1982). The indirect tax effects presented in this section are attributable solely to a general reduction in final demand. In fact they are estimated as if there were no change in the state and local tax structure: This specification is formally built into our estimates by assuming that for each tax analyzed, tax receipts per dollar of sales in each industry remain constant (at current levels) when the severance tax is imposed. In this fashion changes in estimated sales in each industry are used to compute these secondary effects on major revenue sources. The four most important instruments in the state, the individual income tax, corporation income tax, retail sales tax, and local property tax, are examined in this fashion. Details of the methodology and data used to form these estimates are presented in appendix *D*.

Estimates of these tax effects appear in tables 6.15 and 6.16. Losses in individual income and retail sales taxes result primarily from reductions

TABLE 6.15
Reductions in Other Tax Revenues Attributable to Reduced Drilling and Oil
Production: 9-Percent New-Well Decline Rate
(millions of dollars annually)

Revenue Source	10-Year Average	20-Year Average	30-Year Average
Individual income tax	8.2	11.0	12.5
Corporation income tax	9.3	14.2	17.6
Retail sales tax	8.7	11.5	13.0
Local property tax	3.8	5.3	6.4
Total	30.0	42.0	49.5

SOURCE: See text.

TABLE 6.16
Reductions in Other Tax Revenues Attributable to Reduced Drilling and Oil
Production: 3-Percent New-Well Decline Rate
(millions of dollars annually)

Revenue Source	10-Year Average	20-Year Average	30-Year Average
Individual income tax	8.6	12.2	14.6
Corporation income tax	10.1	16.8	22.0
Retail sales tax	9.0	12.7	15.1
Local property tax	4.1	6.2	7.7
Total	31.8	47.9	59.4

SOURCE: See text.

in household income. These impacts are largely attributable to reduced drilling, since direct and indirect labor requirements in drilling are more important than in crude oil production. This mainly accounts for the finding that losses to these two tax instruments remain relatively constant through time. Corporate income tax receipts, on the other hand, are more sensitive to the production effect and thus show a steady decline over the periods examined.

Gross proceeds from the severance tax (without small-producer exemptions) are expected to range from $500 to $560 million per year if real crude oil prices remain at current levels. Consequently the offsets in tables 6.15 and 6.16 represent 5–10 percent of the revenue anticipated from the severance tax.

In addition to the preceding offsets, the proposed severance tax would reduce the base of the state corporate income tax. The amount by which the corporate tax base would fall depends on the effective unitary tax apportionment ratio used to compute corporate tax liabilities in California. This is a subject on which there is little empirical information. Since the state corporate income tax rate is 9.6 percent, these corporate tax offsets would, at most, amount to 9.6 percent of gross severance tax receipts. If severance tax receipts averaged $530 million per year, the implied state corporate tax offset would then be $50.9 million per year (Camm et al. 1982).

Corporate tax offsets associated with simple deductibility are separate from, and additive to, the tax effects presented in tables 6.15 and 6.16. When summed, these estimates indicate that the proposed severance tax would reduce annual collections from other state and local taxes by at most $110.3 for the 30-year horizon.

6.4 Conclusion

A 6-percent severance tax on crude oil production would tend to reduce gross sales and employment demand in California. Employment would fall directly due to reductions in crude production and drilling, and indirectly because of the multiplier effect as initial reductions in activity spread through the California economy. The impact would tend to grow over time, since the production impact on existing wells falls heaviest near the end of the economic lifetimes of those wells. Estimated reductions in employment demand range between 9,000 and 16,000 positions, depending on the forecast period considered and the production decline rate of newly drilled oil wells.

These estimates should be interpreted with caution, however. A reduction in employment demand of 12,000 jobs (the 20-year average, assuming a 9-percent new-well decline rate) does not mean that 12,000 workers will be permanently added to the unemployment rolls. Most workers whose jobs are lost because of the tax-induced reduction in activity will eventually seek other jobs in the state or leave California altogether. Nor will the effects of the tax on employment be felt uniformly throughout the state. As pointed out in section 6.3.2, reductions in employment demand are likely to be concentrated in certain geographic areas and sectors most closely linked with the crude production and drilling industries. Furthermore, economic forecasts of the type presented here necessarily contain a degree of uncertainty, and the fixed coefficients of the input-output model are likely to change as factor prices and structural relationships evolve in the future. No one can predict with certainty such factors, or the course of world crude oil prices, over the next 30 years.

Nevertheless, the forecasts we have computed are not without force. They are based on the best available data on reserves, production, and drilling. The level of analysis is highly disaggregated, to the field level in estimating production effects, and to quite homogeneous sectors in the DWR input-output model. Where critical assumptions were necessary, they were chosen to avoid overestimating effects of the tax.

Given the budgetary difficulties faced by the state of California, it is clear that all levels of government in the state face a period of challenging problems and difficult decisions. We believe that the people of the state will best be served if those responsible for fiscal policy have available the most detailed information and accurate forecasts that are obtainable from economic studies. We hope that our analysis of the effects of the proposed crude oil severance tax will contribute in a useful and informative way to public discussion of the choices facing us.

Appendix D

The Economic Impact Model: Analytical Details

The input-output framework provides a simple representation of economic interrelationships in general equilibrium. The input-output approach is concerned primarily with interrelationships in production. Because, as explained in the text, the allocative effects of the severance tax are expected to be felt exclusively in the production or supply side of the California economy, the input-output approach is a natural vehicle for analyzing these effects. This appendix presents a brief overview of the input-output paradigm, explains how the technique is applied to answer questions pertinent to the evaluation of the severance tax, and gives sources of data required to implement the analysis. The discussion of input-output theory is brief and is intended for readers who are already familiar with the elements of the technique. For those who wish to obtain a basic understanding of input-output analysis, texts by Baumol (1977) and Richardson (1972) are recommended.

D.1 Estimating Economic Impacts with Input-Output Analysis

Input-output analysis proceeds from three key assumptions on the production technology: (1) each industry produces a single homogeneous product, (2) production relationships exhibit fixed coefficients (fixed input-output ratios), (3) returns to scale are constant. The consumption side of the economy is characterized by fixed final demands. To proceed, we adopt the following notation:

$i = 1...n$	An index of industries in the economy
x_i	The amount of output produced by industry i per period
a_{ij}	The input-output coefficient for purchase by industry j from industry i (explained below)
d_i	Final demand for output of industry i

In the following presentation, lower case letters used without subscripts refer to $(n \times 1)$ column vectors, upper case letters refer to $(n \times n)$ matrices, and the term " ' " refers to the transpose operator.

The term a_{ij} is the amount of industry i output required to produce a unit of output of industry j. Let $\mathbf{A}$ refer to the matrix of these coefficients, where the elements $(a_{i1}...a_{in})$ appear in row i. The vector of industry outputs required to support production of a given output vector $\mathbf{x}$ ($\mathbf{x}' = x_1...x_n$) is $\mathbf{Ax}$. If in addition to these interindustry outputs, a vector of final demands $\mathbf{d}$ ($\mathbf{d}' = d_1...d_n$) is to be satisfied, the required output vector may be expressed as

$$\mathbf{x} = \mathbf{Ax} + \mathbf{d}. \tag{D.1}$$

Solving for $\mathbf{x}$ yields

$$\mathbf{x} = (\mathbf{I} - \mathbf{A})^{-1}\mathbf{d} \tag{D.2}$$

where $\mathbf{I}$ is an $(n \times n)$ identity matrix.

Because $\mathbf{x}$ is linear in $\mathbf{d}$, it is relatively easy to solve for changes in industry outputs that would accompany a given change in final demand:

$$\Delta\mathbf{x} = (\mathbf{I} - \mathbf{A})^{-1}\Delta\mathbf{d} \tag{D.3}$$

where Δ denotes a change in a variable and it is assumed that the matrix $\mathbf{A}$ is invariant to changes in $\mathbf{x}$.

If levels of labor per unit output in each industry are known, then equation (D.3) can be used to estimate the effects of a change in final demand on labor requirements, under the assumption that labor per unit output remains fixed. Let l_i denote the amount of labor employed per unit output in industry i and let $\mathbf{L}$ represent a diagonal matrix in which these l_i terms appear along the principal diagonal; all other elements in $\mathbf{L}$ are zero. Finally let $e_i = l_i x_i$ denote total employment in industry i. Then the effects of a change in final demand on labor requirements may be written

$$\Delta\mathbf{e} = \mathbf{L}\Delta\mathbf{x}$$

$$= \mathbf{L}(\mathbf{I} - \mathbf{A})^{-1}\Delta\mathbf{d} \tag{D.4}$$

where $\mathbf{e}' = (e_1...e_n)$.

The approach just outlined can also be used to estimate the effects of a change in final demand on tax receipts. Let the subscript k refer to the *k*th tax instrument (for example, the corporate income tax) and let t_{ki} be the amount of tax paid per unit output by industry *i*. Then $u_{ki} = t_{ki}x_i$ is the amount of tax *k* collected from industry *i*. A matrix T_k of tax coefficients may now be defined, with elements t_{ki} appearing on the principal diagonal and zeros elsewhere. The vector of tax receipts from each industry $(u'_k = u_{kl} \dots u_{kn})$ is expressed

$$u^k = T_k x$$
$$= T_k (I - A)^{-1} d. \qquad (D.5)$$

Under the assumption that T_k and A are fixed, the effect of a change in final demand on tax collections from each industry is

$$\Delta u^k = T_k (I - A)^{-1} \Delta d. \qquad (D.6)$$

The procedure just outlined was carried out both with regard to employment effects of a severance tax and with regard to effects on four major tax revenue sources in California: the individual income tax, the corporate income tax, the retail sales tax, and the local property tax. In terms of the notation already developed, this analysis required the following vectors and matrices: Δd, A, L, and T_k for the four taxes analyzed. In the remainder of this appendix the sources of requisite data are explained in detail.

D.1.1 The Input-Output Model

The input-output model employed in our estimates was prepared by the California Department of Water Resources (DWR) (1980), for use in examining economic impacts of the statewide drought that occurred in the mid 1970s. Interestingly, a primary use of the model by DWR was estimating employment effects of the drought and government responses to the drought (for example, developing alternative water supplies). The model developed by DWR contains 157 sectors, one of which is "households." Thus with households endogenous, the model incorporates induced effects on consumer spending from any shift in final demand.

The DWR input-output table was developed from data pertaining to 1976. The time and resources available for this study did not permit us to update the model's technical coefficients (the A matrix) to reflect possible changes in the production technology since 1976. It is, of course, possible that such changes have occurred. One potential reason for such changes would be the energy price increases that occurred from 1979 to

1981. Our study is primarily designed to examine changes that result from shifts in final demand in energy-related sectors. Despite this, however, it is not expected that such changes would have a significant impact on our estimates. The reason is that these potential changes would appear mainly in purchases of energy *by other* sectors of the economy; the employment and tax effects of the severance tax, on the other hand, are generated by changes in purchases by energy-related industries *from other* sectors of the economy. (In terms of the preceding notation, the energy price increases in the late 1970s would be expected to affect the energy-related *rows* of the **A** matrix. However, our impact estimates are sensitive mainly to the energy-related *columns* of **A**.) As a consequence these possible changes in energy purchases per unit output are no more serious in the present context than they would be for analyzing a change in final demand in any other sector in the model.

D.1.2 Change in Final Demand

As noted in chapter 5, direct effects of the proposed severance tax are expected to arise from changes in production of crude oil, both from existing and new reserves and from reductions in the number of new oil wells drilled. These effects are simulated by introducing appropriate changes in the final demand vector **d** of the input-output model. In the DWR model final demands are entered in dollar amounts with outputs valued at 1976 wholesale or producers prices (California Department of Water Resources 1980, p. 64). To use the model to evaluate current impacts, it is necessary to adjust current dollar changes in final demands to reflect 1976 prices. The resulting dollar impacts, stated in 1976 terms, may then be inflated to 1982 values for easier interpretation. To perform this task it is convenient to define a price adjustment matrix **P**. Let subscripts 0 and 1 refer to the years 1976 and 1981, respectively. The producer price indexes for industry i in periods 0 and 1 are then written p_{i0} and p_{i1}, respectively. The matrix **P** is constructed with price index ratios p_{i0}/p_{i1} along the principal diagonal and zeros elsewhere. With **P** defined in this fashion, the 1976 value of any current change in final demand becomes

$$\Delta d_0 = P \Delta d_1 \qquad\qquad (D.7)$$

where Δd_0 and Δd_1 refer, respectively, to final demand changes expressed in 1976 and 1982 prices. With final demand changes properly expressed in 1976 terms, equation (D.3) can be used to evaluate changes in industry outputs. However, unless prices are readjusted to current terms, these industry output changes will be expressed in 1976 terms. To place them in current dollar values for easier interpretation, they must be inflated by multiplying the output change vector (Δx) by the inverse of **P**. The resulting expression for the change in industry output expressed in current dollar values (Δx_1) is

$$\Delta x_1 = P^{-1}(I - A)^{-1}P\Delta d_1. \tag{D.8}$$

To specify the appropriate change in final demand, it is useful to think of demand as the sum of consumption plus investment plus exports minus imports. Because crude oil prices are assumed to be set in international markets, they will not change as a result of the tax. Correspondingly, total consumption of crude oil, refined products, and so forth, is unaffected by the tax. All changes in final demand, then, are confined to investment and net trade components. In keeping with this specification, reductions in California crude oil output would be exactly replaced by imports (presumably from Alaska or Indonesia). Hence the net effect of this impact can be simulated by simply lowering final demand in the crude oil production sector by an amount corresponding to value of reduced output. Drilling new oil wells, on the other hand, is an investment activity. Impacts that stem from reduced drilling are simulated by reducing final demand in the drilling sector by an amount that corresponds to foregone expenditure on new oil wells. In the DWR model the sectors involved in crude oil production and drilling are, respectively, sector 41 ("crude petroleum") and sector 49 ("new construction, all other").

Any substitution for crude oil produced in California by imports will of course involve some changes in the way crude oil is handled, transported, stored, and so forth. These possible reallocations are ignored here, in part because the resources available for this study did not permit us to investigate them thoroughly. It is, however, highly unlikely that such changes would be a source of significant tax or employment impacts. Most crude oil storage is presently located at refineries; hence a shift in the composition of supply sources is not expected to have a major impact on crude oil storage. With imports substituted for domestic supplies, there would be some shift in modes of transport from pipelines to waterborne tankers. Here any increase in imports from abroad would probably be in foreign-owned vessels and hence not a source of increased final demand for California firms. Increased imports from Alaska would be in U.S. vessels, but even here only a fraction of these would be owned by California firms. Pipeline transport from the oilfield to the refinery, on the other hand, would decrease and this is largely a domestic enterprise. Hence any losses to this sector would be felt primarily within California. Therefore ignoring all such transport reallocations will, if anything, impart conservative bias to our impact estimates.

D.1.3   Employment and Tax Impact Matrices

To construct the **L** matrix it is necessary to have data on employment per unit output by industry. The DWR input-output study constructed these employment coefficients for each industry in the model. To allow for productivity changes, employment coefficients in all industries were adjusted to reflect the change in output per man hour in California since 1976. The data that would be required to make detailed, industry-specific adjustments were unavailable.

To simulate tax impacts it is necessary to estimate taxes paid per unit output. Because requisite data are not available for 1982, we assumed (for all taxes except the property tax) that tax paid per unit output in each industry has not changed since 1976. (The approach adopted for the property tax is explained later.) To estimate taxes paid by sector in 1976, actual tax collections in that year were attributed to individual sectors in the DWR input-output model. All individual income tax payments were attributed to the households sector. Sales and use taxes were assumed to originate in the retail trade sector. Receipts from the corporate income tax were distributed among sectors according to the distribution of profit-type income in the value-added portion of the DWR transactions matrix. These estimates of taxes paid by industry in 1976 were then divided by 1976 industry output levels to obtain the needed tax coefficients.

The nature of property taxation in California has changed dramatically since 1976. To incorporate the fact that proposition 13 froze assessments on much of the taxable property in California, it was assumed that only one-third of the property tax receipts of local governments are sensitive to changes in economic conditions. This portion of property tax revenue was then distributed among sectors according to total profit income and household income as reported in the DWR transaction matrix. Implicit in this apportionment formula is the view that changes in property assessment on business and residential property will primarily be related to changes in the incomes of property owners.

D.2 Data Required to Implement the Model

Definitions of individual industries in the DWR input-output model are reported in table D.1. The 1976 value of output for each industry appears in column 1 of this table, employment is shown in column 2, and employment output coefficients in column 3.

Expenses attributed to drilling wells are reported in table D.2. The total cost per well (1982 dollars) is $638,137. To form this estimate we began with the list of oil extraction expenses reported by the U.S. Bureau of the Census (1982) in its *Annual Survey of Oil and Gas* and wherever possible inserted cost figures specific to California. Where state-specific data were unavailable, averages for the lower 48 were used.

The cost of drilling and equipping wells was taken from American Petroleum Institute (1981) and pertains directly to California. Statewide figures were adjusted to reflect the distribution of onshore versus off-shore wells in tax eligible fields in the state. Data reported by the U.S. Energy Information Administration (1982a, 1982b) were used to update 1980 cost estimates to 1982. Geological and other expenses are nation-wide averages taken from U.S. Bureau of the Census (1982). Reported 1980 figures were inflated by the producer price index for all commodities to reflect 1982 price levels.

Lease equipment costs were estimated separately for producing oil wells, water injection wells, and steam injection wells. It was assumed that the distribution of future oil wells among these categories, and the percent of all wells that are dry holes, will follow the pattern observed for tax eligible fields in 1980. Lease equipment costs for producing wells and for water injection wells were taken from U.S. Energy Information Administration (1982). Equipment costs for steam injection wells were estimated from information reported by Lewin and Associates (1981). Costs for steam injection were assumed to equal costs for water injection plus an estimated $12,500 for steam-generating equipment. The latter figure is based on an estimated 1976 cost of $300,000 for a 50-million BTU-per-hour steam generator and an assumed daily injection rate of 55 bbl. of steam per injection well. (In table D.2 the category "steam injection wells" includes a few wells that are actually used for air injection and pressure maintenance.)

D.3 Other Considerations

To avoid the possibility of double counting in our impact estimates, the list of expenses attributed to drilling wells in table D.2 was compared to

Industries in the DWR Input-Output Model

Industry	Total Production ($ millions)	Total Labor Use (person-years)	Direct Resource Coefficient (person-years per $ million)
1. Dairies	1082.	14516.0	13.41
2. Broilers, chickens and eggs	580.	9992.0	17.24
3. Turkeys and other poultry	124.	1402.0	11.33
4. Cattle and calves	779.	8250.0	10.59
5. Hogs	42.	912.0	21.70
6. Sheep, lambs, and wool	64.	1390.0	21.67
7. Miscellaneous livestock	9.	195.9	21.68
8. Apiary products	19.	299.0	15.50
9. Cotton	833.	10187.0	12.24
10. Wheat	229.	2818.0	12.30
11. Rice	161.	2848.0	17.71
12. Barley	157.	2424.0	15.47
13. Corn	181.	2744.0	15.18
14. Hay and pasture	794.	7891.0	9.94
15. Oats	14.	159.0	11.56
16. Sorghum grain	44.	606.0	13.88
17. Grass seed	35.	281.0	7.92
18. Food, feed grains, not elsewhere classified	0.	4.9	12.30
19. Tobacco	0.	0.0	0.0
20. Walnuts	109.	4006.0	36.71
21. Almonds	182.	5531.0	30.38
22. Noncitrus fruits	1344.	114408.0	85.11
23. Citrus fruits	430.	16302.0	37.89
24. Fruit and tree nuts, not elsewhere classified	1.	94.0	85.11
25. Vegetables	1388.	59183.0	42.64
26. Dried beans	69.	1084.0	15.62
27. Dried peas	0.	0.0	0.0
28. Melons	116.	5203.0	44.97
29. Sugar beets	192.	2935.0	15.27
30. Hops	1.	13.0	8.92
31. Potatoes	94.	1002.0	10.66
32. Sweet potatoes	18.	650.0	36.90
33. Vegetables and sugar, not elsewhere classified	0.	0.0	0.0

Industry	Total Production ($ millions)	Total Labor Use (person-years)	Direct Resource Coefficient (person-years per $ million)
34. Safflower	37.	223.0	6.00
35. Oil crops not elsewhere classified	0.	0.0	0.0
36. Greenhouse and nursery products	506.	40196.0	79.40
37. Forestry and fishery products	376.	4053.0	10.79
38. Agriculture, forestry, fishery services	1298.	129562.0	99.79
39. Metals mining	128.	2292.0	17.86
40. Coal mining	0.	0.0	0.0
41. Crude petroleum	2932.	11876.0	4.05
42. Natural gas and natural gas liquids	333.	2094.0	6.28
43. Stone and clay mining and quarring	411.	6586.0	16.04
44. Chemical and fertilizer mineral mining	145.	1430.0	9.88
45. New construction residential	8862.	163060.0	18.40
46. New construction nonresidential	5834.	77733.0	13.32
47. New construction public utility	3691.	38200.0	10.35
48. New construction highways	1295.	19100.0	14.75
49. New construction all other	2043.	27723.0	13.57
50. Maintenance and repair construction	4559.	76719.0	16.83
51. Ordnance and guided missiles	3663.	38751.0	10.58
52. Meat products	2761.	19865.0	7.19
53. Dairy products	2151.	13389.0	6.22
54. Canned and frozen foods	6951.	56453.0	8.12
55. Grain mill products	1402.	7776.0	5.55
56. Bakery products	1139.	18678.0	16.40
57. Sugar	959.	5260.0	5.49
58. Confectionary products	293.	5081.0	17.36
59. Beverages and flavorings	2951.	21134.0	7.16

Industry	Total Production ($ millions)	Total Labor Use (person-years)	Direct Resource Coefficient (person-years per $ million)
60. Miscellaneous food products	3853.	30005.0	7.79
61. Tobacco manufacturers	0.	0.0	0.0
62. Textile products	4177.	118085.0	28.27
63. Logging camps and sawmills	2046.	27286.0	13.34
64. Millwork, plywood and other wood products	1780.	22353.0	12.56
65. Wooden containers	206.	4748.0	23.06
66. Household furniture	1212.	32856.0	27.10
67. Office furniture and fixtures	706.	16196.0	22.95
68. Paper and paperboard products	2848.	36982.0	12.98
69. Newspapers	1045.	37790.0	36.15
70. Other printing and publishing	2972.	68540.0	23.06
71. Industrial chemicals	1892.	9176.0	4.85
72. Agricultural chemicals	881.	6467.0	7.34
73. Gum and wood chemicals	725.	6646.0	9.16
74. Plastics materials and synthetic fibers	381.	3389.0	8.90
75. Drugs	765.	12090.0	15.80
76. Cleaning and toilet preparations	1410.	12500.0	8.86
77. Paints and allied products	831.	7736.0	9.31
78. Petroleum refining and related products	11026.	25559.0	2.32
79. Rubber and plastics products	2950.	53044.0	17.98
80. Leather tanning and products	287.	10451.0	36.47
81. Glass	809.	15947.0	19.71
82. Cement and concrete products	1135.	16355.0	14.41
83. Structural clay products	126.	3868.0	30.60
84. Pottery and related products	120.	7035.0	58.45
85. Miscellaneous stone and clay products	407.	10131.0	24.91

Industry	Total Production ($ millions)	Total Labor Use (person-years)	Direct Resource Coefficient (person-years per $ million)
86. Blast furnaces and basic steel products	1797.	20948.0	11.66
87. Iron and steel foundries and forgings	415.	8695.0	20.97
88. Primary nonferrous metal products	3161.	23546.0	7.45
89. Metal containers	1106.	13216.0	11.95
90. Heating apparatus and plumbing fixtures	280.	6109.0	21.78
91. Fabricated structural steel	2027.	35596.0	17.56
92. Screw machine products	458.	8739.0	19.09
93. Metal stampings	470.	12059.0	25.68
94. Cutlery, hand tools and general hardware	638.	16188.0	25.37
95. Other fabricated metal products	1892.	30703.0	16.23
96. Engines, turbines and generators	875.	6129.0	7.00
97. Farm machinery	364.	4116.0	11.30
98. Construction and material-handling equipment	1209.	21745.0	17.98
99. Metal-working machinery	1027.	17715.0	17.25
100. Special industrial machinery	854.	10367.0	12.14
101. General industrial machinery	1149.	17808.0	15.50
102. Machine shop products	1089.	29432.0	27.03
103. Computers and office equipment	2817.	63224.0	22.45
104. Service industry machines	512.	5414.0	10.58
105. Electric transmission equipment	917.	8471.0	9.24
106. Electrical industrial apparatus	608.	17199.0	28.28
107. Household appliances	352.	5451.0	15.49
108. Electric lighting and wiring	691.	17463.0	25.28
109. Radio and television receiving sets	654.	16155.0	24.71

Industry	Total Production ($ millions)	Total Labor Use (person-years)	Direct Resource Coefficient (person-years per $ million)
110. Communication equipment	4650.	101492.0	21.83
111. Electronic components	3781.	74584.0	19.72
112. Miscellaneous electorical products	446.	6104.0	13.69
113. Motor vehicles	4572.	40384.0	8.83
114. Aircraft	9081.	119930.0	13.21
115. Ship and boat building and repairing	1050.	22194.0	21.13
116. Other transportation equipment	895.	24296.0	27.14
117. Clocks and scientific equipment	2802.	68465.0	24.43
118. Jewelry, sporting goods, etc.	1494.	40632.0	27.20
119. Railroads	1524.	32716.0	21.46
120. Local transit and intercity buses	613.	25248.0	41.21
121. Truck transportation	3790.	116005.0	30.61
122. Water transportation	1815.	19328.0	10.65
123. Air transportation	4480.	67597.0	15.09
124. Pipeline transportation	71.	438.0	6.15
125. Transportation services	444.	23228.0	52.34
126. Communication except radio and television	5942.	121452.0	20.44
127. Radio and television broadcasting	845.	19269.0	22.81
128. Electric companies and systems	3803.	22220.0	5.84
129. Gas companies and systems	3335.	18720.0	5.61
130. Water and sanitary services	1228.	22928.0	18.67
131. Wholesale trade	12214.	525674.0	43.04
132. Retail trade	24255.	1602181.0	66.06
133. Banking and financial intermediaries	5233.	235547.0	45.02
134. Insurance	6340.	151162.0	23.84
135. Owner-occupied real estate	11956.	0.0	0.0
136. Real estate	13825.	129660.0	9.38
137. Hotels and lodging places	1309.	137725.0	105.21

Industry	Total Production ($ millions)	Total Labor Use (person-years)	Direct Resource Coefficient (person-years per $ million)
138. Personal and repair services	2770.	182110.0	65.75
139. Miscellaneous business services	10840.	338461.0	31.22
140. Advertising	4649.	16973.0	3.65
141. Miscellaneous professional services	5909.	175234.0	29.65
142. Automobile repair	3303.	86277.0	26.12
143. Motion pictures	4530.	66929.0	14.78
144. Amusement and recreation services	2290.	82738.0	36.13
145. Doctors and dentists	5074.	164801.0	32.48
146. Hospitals	4075.	206787.0	50.74
147. Other medical services	2333.	142920.0	61.25
148. Educational services	1881.	106288.0	56.51
149. Nonprofit organizations	2139.	133343.0	62.33
150. Post office	1153.	68925.0	59.78
151. Oher federal government enterprises	381.	2590.0	6.80
152. State and local government enterprises	2032.	59959.0	29.51
153. Noncompetitive imports	0.	0.0	0.0
154. Dummy industries	3665.	0.0	0.0
155. Government industry	20908.	1532855.2	73.31
156. Special industries	0.	0.0	0.0
157. Household income	116193.	0.0	0.0

the pattern of direct purchases by the corresponding industry (sector 49, new construction, all other) in the DWR input-output model. Inspecting the DWR transactions matrix revealed that none of the drilling expenses are included among purchases in the crude oil production sector. (That is, sector 41 does not purchase from sector 49 in the DWR model; rather, sales by sector 49 are directly to the investment component of final demand.) Thus it is appropriate to enter crude oil production and oil well drilling expenses in an additive fashion when simulating employment and tax impacts.

Comparing individual expense items and the DWR transactions table did reveal one important consideration, however. As reported by the U.S. Bureau of the Census (1982), the acquisition of land (both producing and nonproducing acreage) is a major oil industry expense. In the Bureau of

TABLE D.2
Expenses Attributable to Drilling Oil Wells
(1982 dollars per well)

	Percent in Sample[a]	Cost of Drilling and Equipping[b]	Leased Equipment[c]	Geological and Other[d]
Producing wells	64.52	—	$105,431	—
Water injection wells	6.79	—	$ 98,630	—
Steam injection wells	18.53	—	$111,130	—
Dry holes	10.16	—	—	—
Weighted average	100.00	$297,680	$ 95,313	$245,143

SOURCES:

[a] U.S. Bureau of the Census, *Annual Survey of Oil and Gas* (Washington, D.C.: Department of Commerce, 1981).

[b] American Petroleum Institute, *1980 Joint Association Survey of Drilling Costs* (Washington, D.C.: 1981).

[c] U.S. Energy Information Administration, *Costs and Indexes for Domestic Oil and Gas Field Equipment and Production Operations, 1981* (Washington, D.C.: Department of Energy, 1982a).

[d] U.S. Energy Information Administration, *Indexes and Estimates of Domestic Well Drilling Costs, 1981 and 1982,* (Washington, D.C.: Department of Energy, 1982).

NOTE: Percent in sample pertains to tax eligible fields for which production impacts were estimated. Cost of drilling and equipping wells includes only equipment through the "Christmas tree" on producing wells. Geological and other includes costs of exploratory crews, scouting, general administration and direct overhead.

the Census classification, this item is attributed to exploration and development rather than production. Thus it might seem appropriate to include this cost item as a component of the final demand for drilling. However, having inspected the DWR transactions matrix, it appears that land acquisitions (primarily royalty payments) are considered purchases by the crude oil production industry (sector 41) within the input-output model. Hence in computing production and drilling impacts, it was assumed that all land purchases and lease payment expenditures are captured by final demand changes in the crude production sector.

Finally, it is worth emphasizing that our estimates of employment and tax impacts follow solely from reduction in drilling and crude oil production. These are in no way modified by, or specific to, the assumption that the underlying production effects arise from a new severance tax. That is, our estimated employment and tax impacts would be no different if crude oil production and drilling reductions were attributable to some other cause. In addition to reducing production and drilling in the state, an increase in the severance tax would also transfer a substantial amount of purchasing power away from the industry and to the government. This transfer is of course simply the severance tax payment on crude oil that would continue to be produced after imposition of the tax. Increased tax revenue would indeed allow government to spend more. At the same time, however, an equivalent reduction in the industry's spendable income would occur. (Within the input-output paradigm this reduction would occur in the value-added sectors of the model—profits, interest, taxes, and depreciation.) With the analytical methods now available, it is not possible to determine the impact of such a transfer of purchasing power on economic activity. As a consequence, these two offsetting effects are not examined.

Bibliography

Advisory Commission on Intergovernmental Relations. *Significant Features of Fiscal Federalism, 1980–1981 Edition.* Washington, D.C., 1981.

Alaska Department of Revenue, *Revenue Sources FY 1982–83* (Juneau, Alaska: January 1983).

Alaska State Assessor's Office, *Alaska Taxable Property,* various years (Juneau, Alaska).

American Petroleum Institute. *1980 Joint Association Survey on Drilling Costs.* Washington, D.C., December 1981.

———. *Basic Petroleum Data Book.* Washington, D.C.: 1982.

———. *Quarterly Review of Drilling Statistics for the United States* (various issues). Washington, D.C.

———. *Reserves of Crude Oil, Natural Gas Liquids, and Natural Gas in the United States and Canada as of December 31, 1979.* Washington, D.C.: June 1980.

Arps, J. J. and T. G. Roberts. "Economics of Drilling for Cretaceous Oil on East Flank of Denver-Julesberg Basin." *Bulletin of the American Association of Petroleum Geologists* 42, no. 11 (November 1958).

Attanasi, Emil D., "Economics and Resource Appraisal—The Case of the Permian basin." *Journal of Petroleum Technology,* April 1981, pp. 603–616.

———, L. J. Drew, and D. H. Root. "Physical Variables and the Petroleum Discovery Process." In Ramsey, James, ed. *The Economics of Exploration for Energy Resources.* JAI Press, 1981.

Baumol, William J. *Economic Theory and Operations Analysis.* Englewood Cliffs: Prentice Hall, 1977.

Belal, Roshida. "Windfall Profit Tax, 1982." *Statistics of Income Bulletin* (U.S. Internal Revenue Service) 2, no. 1 (Summer 1982).

Cal-Tax Research. *Bulletin October 1982.* Sacramento, Calif.: 1982.

California Assembly Office of Research. *State Taxation on the Production of Crude Oil: A Comparison of Nine States.* Sacramento, Calif.: California State Legislature, 1981.

California Department of Water Resources. *Measuring Economic Impacts: The Application of Input-Output Analysis to California Water Resources Problems.* Bulletin 210. Sacramento, Calif.: Resources Agency, 1980.

California Division of Oil and Gas. *Annual Report of the State Supervisor of Oil and Gas* (various issues). Sacramento, Calif.

California Employment Development Dept. *Annual Planning Information 1981– 1982*. Sacramento, Calif.: May 1981.

———. *California Labor Market Bulletin, Statistical Supplement*. Sacramento, Calif.: December 1976.

———. *California Labor Market Bulletin, Statistical Supplement*. Sacramento, Calif.: February 1981.

California Franchise Tax Board. *Annual Report* (various issues). Sacramento, Calif.

———. "Overview of State Taxation of Gas and Oil Production Income." Prepared for Assembly Revenue and Taxation Committee, interim hearings. Monterey, Calif.: 21 and 23 September 1982.

California Legislative Analyst. Letter from Wm. G. Hamm to Hon. Wadie P. Deddeh, Assembly Eightieth District. Sacramento, Calif.: 10 September 1982a.

———. Letter from Wm. G. Hamm to Hon. John Vasconcellos, Chairman Assembly Committee on Ways and Means. Sacramento, Calif.: 20 April 1982(b).

California Office of Planning and Research. *Economic Practices Manual, a Handbook for Preparing an Economic Impact Assessment*. Sacramento, Calif.: 1978.

California State Board of Equalization. *County Oil and Gas Survey*. Sacramento, Calif.: 1981.

California State Comptroller. *Annual Report 1980–81* (various issues). Sacramento, Calif.

Camm, Frank, C. W. Myers, R. Y. Arguden, S. J. Bell, and T. Jacobsson. *Effects of a Severance Tax on Oil Produced in California*. Prepared for the California State Assembly, Rand Corporation. Santa Monica, Calif.: 1982.

Commonwealth Edison versus Montana. 101 S. Ct. 2946 (1981).

Complete Auto Body versus Brady. 430 U.S. 274 (1977).

Conservation Committee of California Oil Producers. *Annual Review of California Oil and Gas Production*. Los Angeles, Calif.: 1982.

———. *California Annual and Cumulative Crude Oil Production-Barrels by Fields and Pools 1930–1970*. 1971.

Cox, James C., and Arthur W. Wright. "The Determinants of Investment in Petroleum Reserves and their Implications for Public Policy." *American Economic Review* 66, no. 1 (1976): 153–67.

David, Lee B. "Thirty-Percent Coal Severance Tax Does Not Overburden Interstate Commerce." *Tulane Law Review* 56, no. 4 (June 1982): 1454–67.

Davis, John C., and John W. Harbaugh. "A Simulation Model for Oil Exploration Policy on Federal Lands of the U.S. Outer Continental Shelves." In *The Economics of Exploration for Energy Resources*, edited by James Ramsey. JAI Press, 1981.

Deacon, Robert T., Stephen J. DeCanio, Thomas F. Cooley, H. E. Frech III, and M. Bruce Johnson. *The Proposed California Crude Oil Severance Tax: An Economic Analysis*. The Economics Group, Inc.: Santa Barbara, Calif., 1983.

Epple, Dennis. *Petroleum Discoveries and Government Policy: An Econometric Study of Supply*. Boston: Ballinger Press, 1975.

Epple, Dennis. "The Econometrics of Exhaustible Resource: A Theory and an

Application." *Energy Foresight and Strategy,* edited by Thomas Sargent, Resources for the Future, Inc., Washington, D.C., 1985.

Erickson, Edward W. "Crude Oil Prices, Drilling Incentives, and the Supply of New Discoveries." *Natural Resources Journal* 10, no. 1 (1970): 27–52.

―――, Stephen W. Millsaps, and Robert M. Spann. "Oil Supply and Tax Incentives." *Brookings Papers on Economic Activity* 2 (1974): 449–93.

―――, and Robert M. Spann. "Supply Response in a Regulated Industry: The Case of Natural Gas." *Bell Journal of Economics and Management Science* 2, no. 1 (Spring 1971): 94–121.

Fish, Jerry R. "Commonwealth Edison Co. v. Montana: Leading the Severance Tax Stampede." *Environmental Law* 12, no. 4 (Summer 1982): 1031–58.

Gerking, Shelby D., and John H. Mutti. "Possibilities for the Exportation of Production Taxes: A General Equilibrium Analysis." *Journal of Public Economics* 16, no. 2 (October 1981), 233–52.

Goldberg, Samuel. *Difference Equations.* New York: Wiley, 1958.

Gordon, Roger H. "An Optimal Taxation Approach to Fiscal Federalism." *Quarterly Journal of Economics* 97, no. 4 (November 1973): 567–86.

Independent Petroleum Association of America. *The Oil-Producing Industry in Your State* (various issues). Washington, D.C.

Interstate Oil Compact Commission. *The Effects of the Crude Oil Windfall Profits Tax on Recoverable Crude Oil Resources.* Oklahoma City, Okla.: 1980.

Kalter, Robert J., Thomas H. Stevens, and Oren A. Bloom. "The Economics of Outer Continental Shelf Leasing." *American Journal of Agricultural Economics* 57, no. 2 (May 1975): 251–58.

Khazzoom, J. D. "The FPC Staff's Econometric Model of Natural Gas Supply in the U.S." *Bell Journal of Economics and Management Science* 2 (1971): 51–93.

Kern County Assessor. Testimony presented to Assembly Revenue and Taxation Committee, interim hearings. Monterey, Calif.: 22 and 23 September 1982.

Kim, Young Y., and Russell Thompson. *An Economic Model—New Oil and Gas Supplies in the Lower 48 States.* Ballinger Publishing, 1975.

Lewin and Associates Inc. *Economics of Enhanced Oil Recovery, Final Report.* Washington, D.C.: U.S. Department of Energy, 1981.

Louisiana Department of Revenue and Taxation. *Annual Report* (various issues). Baton Rouge, La.

Louisiana Division of Administration. *State of Louisiana, Annual Financial Report* (various issues). Baton Rouge, La.

Louisiana Tax Commission. *Annual Report* (various issues). Baton Rouge, La.

―――. *Twentieth Biennial Report.* Baton Rouge, La.: 1982.

MacAvoy, Paul W., and Robert S. Pindyck. "Alternative Regulatory Policies for Dealing with the Natural Gas Shortage." *Bell Journal of Economics and Management Science* 4, no. 2 (Autumn 1973), 454–98.

Mancke, Richard B. "The Long-Run Supply Curve of Crude Oil Produced in the United States." *Antitrust Bulletin* 15 (Winter 1970): 727–56.

McClure, Charles E., Jr. "Tax Exporting in the United States: Estimates for 1962." *National Tax Journal* 20, no. 1 (March 1967): 49–78.

―――. "The Interregional Incidence of General Regional Taxes." *Public Finance* 24, no. 3 (1969) 457–83.

―――. "Fiscal Federalism and the Taxation of Economic Rents." In *State and*

Local Finance: The Pressures of the 1980s, ed. by George F. Break. Madison: University of Wisconsin Press, 1984. Pp. 133–60.

Morgan, William E., and John H. Mutti. "The Exportation of State and Local Taxes in a Multilateral Framework: The Case of Business Type Taxes." *National Tax Journal* 37, no. 2 (June 1985): 191–208.

National Petroleum Council. *Enhanced Oil Recovery.* Washington, D.C.: December 1976.

Neal, Randall L., and Margaret E. Reed. "Taxes Can Cause up to 93 Percent Variance in Cash Flow." *World Oil* 195, no. 2 (August 1982): 83–92.

Neri, John A. "An Evaluation of Two Alternative Supply Models of Natural Gas." *Bell Journal of Economics* 8, no. 1 (Spring 1977), 289–302.

Oil and Gas Journal 8 no. 14 (5 April 1982) 139–46.

Oklahoma Tax Commission. *Annual Report* (various issues). Oklahoma City, Okla.

———. *Annual Report for the Fiscal Year Ending June 30, 1981.* Oklahoma City, Okla.: 1982.

Olson, Dennis O. "The Interregional Incidence of Energy Production Taxes." *International Regional Science Review* 9, no. 3 (November 1984): 109–24.

Phelps, Charles E., and Rodney T. Smith. *Petroleum Regulation, the False Dilemma of Decontrol.* Santa Monica, Calif.: Rand Corp., 1977.

Pindyck, Robert S. "The Regulatory Implications of Three Alternative Econometric Supply Models of Natural Gas." *Bell Journal of Economics and Management Science* 5, no. 2 (Autumn 1974), 633–45.

Richardson, Harry W., *Input-Output and Regional Economics.* New York: Wiley, 1972.

Sandler, Todd M., and Robert B. Shelton. "Fiscal Federalism, Spillovers, and the Export of Taxes." *Kyklos* 25, no. 4 (1972) 736–53.

Shepherd, Mark. "Commonwealth Edison Co. v. Montana." *Ecology Law Quarterly* 10, no. 1 (1982): 97–111.

Texas Comptroller of Public Accounts. *Annual Financial Report, 1981, State of Texas.* Austin, Tex.

———. *Annual Report, 1980–81.* Austin, Tex.

Tyner, Wallace E., and Kalter, Robert. *Western Coal: Promise or Problem.* Lexington, Mass.: Lexington Books, 1978.

Uhler, Russell S. "Costs and Supply in Petroleum Exploration: The Case of Alberta." *Canadian Journal of Economics* 9, no. 1 (1976): 72–90.

University of California Los Angeles Business Forecasting Project. *The UCLA Business Forecast for California.* Los Angeles: University of California Los Angeles Graduate School of Management, 1982.

U.S. Bureau of the Census. *Annual Survey of Oil and Gas.* Washington, D.C.: Department of Commerce, 1982.

U.S. Bureau of Economic Analysis. *Definitions and Conventions of the 1972 Input-Output Study.* Washington, D.C. Department of Commerce, 1980.

———. *Survey of Current Business, July 1982.* Washington, D.C.: Department of Commerce, 1982.

U.S. Bureau of Labor Statistics. *Geographic Profile of Employment and Unemployment, 1980.* Washington, D.C.: Department of Labor, 1982.

———. *Industry Wage Survey: Oil and Gas Extraction September 1977.* Washington, D.C.: Department of Labor, February 1979.

———. *Supplement to Employment and Earnings, States and Areas; Data for 1977–1980*. Washington, D.C.: Department of Labor, 1982.

U.S. Cabinet Task Force on Oil Import Control. *The Oil Import Question, Appendix D*. Washington, D.C.: 1970.

U.S. Department of Commerce, Bureau of Economic Analysis. *Survey of Current Business*. (Various issues). Washington, D.C..

U.S. Energy Information Administration. *Applied Analysis Modeling Capability Survey, "Models Index."* Washington, D.C.: Department of Energy, February 1978.

———. *Costs and Indexes for Domestic Oil and Gas Field Equipment and Production Operations* (various issues). Washington, D.C.: Department of Energy, 1982a.

———. *Energy Taxation: An Analysis of Selected Taxes*. Washington, D.C.: Department of Energy, 1980.

———. *Historical Review of Domestic Oil and Gas Exploratory Activity*. Washington, D.C.: Department of Energy, October 1979.

———. *Indexes and Estimates of Domestic Well Drilling Costs, 1981 and 1982*. Washington, D.C.: Department of Energy, August 1982b.

———. *Medium-Run Oil and Gas Supply Model 1977 Data Update*. Washington, D.C.: Department of Energy, December 1977.

———. *Oil and Gas Supply Curves for the Administrator's Annual Report*. Technical memorandum (DDE/EIA-0103/4). Washington, D.C.: Department of Energy, 1978.

———. *Price Controls and International Petroleum Product Prices*. Washington, D.C.: Department of Energy, February 1980.

———. *Production of Onshore Lower-48 Oil and Gas Model Methodology and Data Description*. Washington, D.C.: Department of Energy, June 1982.

———. *The Integrating Model of the Project Independence Evaluation System*. Washington, D.C.: Department of Energy.

———. *U.S. Crude Oil, Natural Gas, and Natural Gas Liquids Reserves* (various issues). Washington, D.C.: Department of Energy.

U.S. Geologic Survey. *Future Supply of Oil and Gas from the Permian Basin of West Texas and Southeastern New Mexico*. Circular 828. Washington D.C.: Department of Interior, 1980.

———. *Outer Continental Shelf Statistics*. 1979–81. Washington, D.C.: Department of Interior, 1982.

U.S. Internal Revenue Service. *Corporation Income Tax Returns, 1978–79*. Washington, D.C.: Department of Treasury, 1980.

———. *Statistics of Income* 2, no. 1 (Summer 1982).

U.S. Minerals Management Service. *Royalties*. Washington, D.C., Department of the Interior, 1982.

Ventura County Assessor. Testimony presented to Assembly Revenue and Taxation Committee, interim hearings. Monterey, Calif.: 22 and 23 September, 1982.

White, L.P. "A Play Approach to Hydrocarbon Resource Assessment and Evaluation." In *The Economics of Exploration for Energy Resources*, edited by James Ramsey. JAI Press, 1981.

Wyoming Department of Administration and Fiscal Control, *Wyoming Data Handbook, 1981* (Cheyenne, Wyo.: 1981).

Wyoming League of Women Voters. *Mineral Wealth and Wyoming Government.* Cheyenne, Wyo.: 1981.
Wyoming Legislative Services Office, Management Council. *Taxes on Wyoming's Minerals: History and Projections.* Cheyenne, Wyo.: September 1982.
Wyoming State Department of Economic Planning and Development. *1981 Wyoming Minerals Yearbook.* Cheyenne, Wyo.: 1981.

Index

Page numbers in italics refer to tables.

Alaska: computing tax burdens, estimation methods, 25–27; effective average tax rates, 14; corporate income taxes, 8, 9–11; data sources for computations, 25–27; property taxes on oil and gas, *26;* receipts from crude oil production, *25;* revenue sources, 5

Arps, 51, 53

Attanasi, geologic-engineering models, 51–53

Blackmun, Harry (U.S. Justice), 42

Brady, Complete Auto Body v., 42

Business taxes, 46, *47*

California, 49; company ownership, 12; computing tax burdens, estimation methods, 27–28; corporate income taxes, 8, 9–11; proposals to raise severance tax on crude oil, 1; receipts from crude oil production, *27;* Department of Water Resources (*see* Department of Water Resources input-output model); anticipated losses of direct jobs, *131;* drilling activity, *110,* 110–11; gravity changes of production, 78; production costs, 17–18; property tax levies, 6, 9; revenue structure, 1–2, 9, 23; royalty collections, 21–22; royalty production,

29; severance tax effects, 69–71; severance tax effects by field, 98, *100–109;* severance tax effects, supply models used to determine, 57–58. *See also* Severance tax, effects on crude oil production; enhanced recovery by steam injection, 19–20; *specific topics,* e.g., Employment

California Assembly Office of Research, 6

California Legislative Analyst, 6

Camm, 53

Commonwealth Edison v. Montana, 41

Complete Auto Body v. Brady, 42

Computing tax burdens: Alaska, 25–27; California, 27–28; corporate income and franchise taxes, 33–38; crude oil, quantity and value of, 38; data sources and methods of estimation, 24–40; individual states, 24–33; Louisiana, 28–30; royalty production, estimated income taxes from, 38–40

Corporate income, estimated tax exportation of, *47*

Corporate income and franchise taxes: computing tax burdens, 33–38; oil production, *36;* tax exempt production from federal lands, *37*

Corporate income tax, 5, 6, 8, *9,* 10, 12, 135